Review of Systemization of the Tooele Chemical Agent Disposal Facility

Committee on Review and Evaluation of the Army
Chemical Stockpile Disposal Program

Board on Army Science and Technology

Commission on Engineering and Technical Systems

National Research Council

NATIONAL ACADEMY PRESS
Washington, D.C. 1996

NOTICE: The project that is the subject of this report was approved by the Governing Board of the National Research Council, whose members are drawn from the councils of the National Academy of Sciences, the National Academy of Engineering, and the Institute of Medicine. The members of the committee responsible for the report were chosen for their special competencies and with regard for appropriate balance.

This report has been reviewed by a group other than the authors according to procedures approved by a Report Review Committee consisting of members of the National Academy of Sciences, the National Academy of Engineering, and the Institute of Medicine.

The National Academy of Sciences, is a private, nonprofit, self-perpetuating society of distinguished scholars engaged in scientific and engineering research, dedicated to the furtherance of science and technology and to their use for the general welfare. Upon the authority of the charter granted to it by the Congress in 1863, the Academy has a mandate that requires it to advise the federal government on scientific and technical matters. Dr. Bruce M. Alberts is president of the National Academy of Sciences.

The National Academy of Engineering was established in 1964, under the charter of the National Academy of Sciences, as a parallel organization of outstanding engineers. It is autonomous in its administration and in the selection of its members, sharing with the National Academy of Sciences the responsibility for advising the federal government. The National Academy of Engineering also sponsors engineering programs aimed at meeting national needs, encourages education and research, and recognizes the superior achievements of engineers. Dr. Harold Liebowitz is president of the National Academy of Engineering.

The Institute of Medicine was established in 1970 by the National Academy of Sciences to secure the services of eminent members of appropriate professions in the examination of policy matters pertaining to the health of the public. The Institute acts under the responsibility given to the National Academy of Sciences by its congressional charter to be an adviser to the federal government and, upon its own initiative, to identify issues of medical care, research, and education. Dr. Kenneth I. Shine is president of the Institute of Medicine.

The National Research Council was organized by the National Academy of Sciences in 1916 to associate the broad community of science and technology with the Academy's purposes of furthering knowledge and advising the federal government. Functioning in accordance with general policies determined by the Academy, the Council has become the principal operating agency of both the National Academy of Sciences and the National Academy of Engineering in providing services to the government, the public, and the scientific and engineering communities. The council is administered jointly by both Academies and the Institute of Medicine. Dr. Bruce M. Alberts and Dr. Harold Liebowitz are chairman and vice chairman, respectively, of the National Research Council.

This is a report of work supported by Contract DAAH04-96-C-0016 between the U.S. Department of the Army and the National Academy of Sciences.

Library of Congress Catalog Card Number 96-67977
International Standard Book Number 0-309-05486-9

Copies available from:
National Academy Press
2101 Constitution Avenue, N.W.
Box 285
Washington, D.C. 20005
(800) 624-6242, (202) 334-3313 (in the Washington Metropolitan Area)

Copyright 1996 by the National Academy of Sciences. All rights reserved.

Printed in the United States of America.

COMMITTEE ON REVIEW AND EVALUATION OF THE ARMY CHEMICAL STOCKPILE DISPOSAL PROGRAM

RICHARD S. MAGEE, *Chair,* New Jersey Institute of Technology, Newark
ELISABETH M. DRAKE, *Vice Chair,* Massachusetts Institute of Technology, Cambridge
DENNIS C. BLEY, Buttonwood Consulting, Inc., Oakton, Virginia
COLIN G. DRURY, University at Buffalo, State University of New York
GENE H. DYER, Consultant, San Rafael, California
MG VINCENT E. FALTER, U.S. Army (Retired), Springfield, Virginia
ANN FISHER, The Pennsylvania State University, University Park
J. ROBERT GIBSON, DuPont Agricultural Products, Wilmington, Delaware
CHARLES E. KOLB, Aerodyne Research, Inc., Billerica, Massachusetts
DAVID S. KOSSON, Rutgers—The State University, Piscataway, New Jersey
WALTER G. MAY, University of Illinois at Urbana–Champaign
ALVIN H. MUSHKATEL, Arizona State University, Tempe
PETER J. NIEMIEC, Greenberg, Glusker, Fields, Claman & Machtinger, Los Angeles, California
GEORGE W. PARSHALL, E.I. du Pont de Nemours & Company, Wilmington, Delaware
JAMES R. WILD, Texas A&M University, College Station
JYA-SYIN WU, Advanced System Concepts Associates, Inc., El Segundo, California (August 1995)

Staff

DONALD L. SIEBENALER, Study Director
MARGO L. FRANCESCO, Administrative Supervisor
SHIREL R. SMITH, Senior Project Assistant
DEBORAH B. RANDALL, Senior Secretary/Project Assistant

BOARD ON ARMY SCIENCE AND TECHNOLOGY

GENERAL GLENN K. OTIS, *Chair*, U.S. Army (Retired), Newport News, Virginia
CHRISTOPHER C. GREEN, *Vice Chair*, General Motors Corporation, Warren, Michigan
ROBERT A. BEAUDET, University of Southern California, Los Angeles
GARY L. BORMAN, University of Wisconsin, Madison
ALBERTO COLL, U.S. Naval War College, Newport, Rhode Island
LAWRENCE J. DELANEY, BDM Europe, Berlin, Germany
JAMES L. FLANAGAN, Center for Computer Aids in Industrial Productivity, Rutgers University, Piscataway, New Jersey
GENERAL WILLIAM H. FORSTER, U.S. Army (Retired), Westinghouse Electronics Systems, Baltimore, Maryland
ROBERT J. HEASTON, Guidance and Control Information Analysis Center, Chicago
THOMAS MCNAUGHER, RAND, Washington, D.C.
NORMAN F. PARKER, Varian Associates (Retired), Cardiff by the Sea, California
STEWART D. PERSONICK, Bell Communications Research, Inc., Morristown, New Jersey
KATHLEEN J. ROBERTSON, Booz, Allen and Hamilton, McLean, Virginia
JAY P. SANFORD, University of Southwestern Health Sciences Center, Dallas, Texas
HARVEY W. SCHADLER, General Electric, Schenectady, New York
JOYCE L. SHIELDS, Hay Management Consultants, Washington, D.C.
CLARENCE G. THORNTON, Army Research Laboratories (Retired), Colts Neck, New Jersey
JOHN D. VENABLES, Martin Marietta Laboratories (Retired), Towson, Maryland
ALLEN C. WARD, University of Michigan, Ann Arbor

Staff

BRUCE A. BRAUN, Director
E. VINCENT HOLAHAN, Senior Program Officer
ROBERT J. LOVE, Senior Program Officer
DONALD L. SIEBENALER, Senior Program Officer
PATRICIA A. KIRCHNER, Administrative Associate
MARGO L. FRANCESCO, Administrative Supervisor
JACQUELINE CAMPBELL-JOHNSON, Senior Project Assistant
ALVERA GIRCYS, Senior Project Assistant
SHIREL R. SMITH, Senior Project Assistant
DEBORAH B. RANDALL, Senior Secretary/Project Assistant

Preface

The United States has maintained a stockpile of highly toxic chemical agents and munitions for more than half a century. In 1985, Congress, in Public Law 99-145, directed the Department of Defense to destroy at least 90 percent of the unitary chemical agent and munitions stockpile, with particular attention to M55 rockets, which were deteriorating and becoming increasingly hazardous. After setting several intermediate goals and dates, Congress, in the National Defense Authorization Act for fiscal year 1993 (P.L. 102-484), dated October 23, 1992, directed the Army to dispose of the entire unitary chemical warfare agent and munitions stockpile by December 31, 2004.

In the 1970s, the Army had commissioned studies of different disposal technologies and tested several of them. In 1982, incineration was selected as the method of disposing of agents and associated propellants and explosives and of thermally decontaminating metal parts. In 1984, the National Research Council (NRC) Committee on Demilitarizing Chemical Munitions and Agents reviewed a range of disposal technologies and endorsed the Army's selection of incineration.

Incineration technology is embodied in today's baseline incineration system, which was developed largely at the Chemical Agent Munitions Disposal System (CAMDS) experimental facility at Tooele Army Depot, Utah. The first full-scale operational plant, the Johnston Atoll Chemical Agent Disposal System (JACADS), is now in service on Johnston Island in the Pacific Ocean southwest of Hawaii. Also, a second plant, the Tooele Chemical Agent Disposal Facility (TOCDF), has been constructed at Tooele Army Depot and has recently undergone systemization (operational testing prior to the start of agent operations), using surrogates for agent to verify that the system and all components will work as designed.

The Committee on Review and Evaluation of the Army Chemical Stockpile Disposal Program (Stockpile Committee) was formed in 1987 at the request of the Undersecretary of the Army to monitor the Army's Chemical Stockpile Disposal Program (CSDP) and to review and comment on relevant technical issues. The Stockpile Committee is a standing committee, which will remain in service with rotating personnel until completion of the disposal program. The committee has monitored the development and implementation of the baseline system and has visited CAMDS numerous times, JACADS three times, and the TOCDF four times. The committee has also reviewed many reports and considerable technical information prepared by the government, government contractors, other agencies, interested civilian groups, and concerned individuals.

In 1993, the Stockpile Committee issued a letter report to the Assistant Secretary of the Army for Installations, Logistics and Environment recommending specific actions to further enhance the Chemical Stockpile Disposal Program risk management process. In early 1994, the Stockpile Committee issued three major reports that included recommendations to the Army concerning changes or improvements to be made to the TOCDF prior to the start of agent operations. These reports are:

- *Evaluation of the Johnston Atoll Chemical Agent Disposal System Operational Verification Testing: Part II.* (Part I was a short summary report issued in July 1993.)
- *Review of Monitoring Activities Within the Army Chemical Stockpile Disposal Program.*
- *Recommendations for the Disposal of Chemical Agents and Munitions.*

The present report continues the work of the four earlier reports by (1) addressing the completion of testing of certain secondary systems that had not been completely tested at JACADS, (2) reviewing the changes implemented by the Army in response to the Stockpile Committee's earlier recommendations pertaining to the Tooele Chemical Agent Disposal Facility, and (3) providing an overview of the status of the facility at the end of the systemization period. This overview is based on the Stockpile Committee's prior knowledge about the baseline system, on information provided by the Army and others, and on site visits to the TOCDF, beginning in October 1991 (midway through the construction phase) through June 1995 (in the late stages of systemization).

The committee greatly appreciates the assistance in support of committee activities and in the production of this report provided by NRC staff members Donald Siebenaler, Margo Francesco, and Deborah Randall; consultants Harrison Pannella and William Spindell; and temporary assistant Julie Harlan.

Richard S. Magee, *Chair*
Elisabeth M. Drake, *Vice Chair*
Committee on Review and Evaluation of the
Army Chemical Stockpile Disposal Program

Contents

EXECUTIVE SUMMARY 1

1 INTRODUCTION 8
   Chemical Stockpile Disposal Program, 8
      The Unitary Chemical Agent and Munitions Stockpile, 8
      Fundamentals of Disposal, 8
      The Baseline Incineration System, 10
   Summary, 10

2 CHANGES AT THE TOOELE CHEMICAL AGENT DISPOSAL
   FACILITY IN RESPONSE TO NATIONAL RESEARCH
   COUNCIL RECOMMENDATIONS 11
   Review of NRC Recommendations Regarding Operational
      Verification Testing at JACADS, 11
         Brine Reduction Area, 11
         Dunnage Furnace, 13
         Nitrogen Oxide Emissions, 15
         Liquid Incinerator Slag Removal, 16
         Furnace Feed System, 17
         Residual Gelled Agent, 18
         Environmental Permitting and Regulatory Requirements, 19
         Environmental Compliance, 19
         Overall Safety, 20
         Changes Resulting from Risk Assessment, 21
   Review of NRC Recommendations Regarding the Monitoring System at JACADS, 21
         General Recommendations for Agent/Nonagent Monitoring, 22
         Specific Recommendations for Agent/Nonagent Monitoring, 23
         Specific Recommendations for Laboratory Operations, 25
         Summary of Responses to Monitoring Recommendations, 26
   Recommendation on Carbon Filtration, 26

3 EVALUATION OF SYSTEMIZATION SAFETY PERFORMANCE 28
   Safety-Related Functions and Reviews by Others, 28
      Systems Hazard Analysis, 28
      Utah Department of Environmental Quality: *Required Report for the Systems
         Hazard Analysis*, 29
      Facility Construction Certification, 30
      Inspector General Report: *Courtesy Chemical Surety Inspection—Tooele CDF*, 32
      TOCDF Safety Evaluation Report, 33
      U.S. Army Chief of Engineers Report: *TOCDF Report on Design-Related Safety
         Issues and Evaluation of Construction Conformance with Design*, 34

   U.S. Army "Lessons Learned" Programs, 34
   U.S. Army Subject Area Review Reports, 35
   State of Utah Inspections, 36
  Stockpile Committee Site Visits, 36
   Personnel Issues (Recruitment, Training, Turnover), 37
   A General Observation, 37
   Shift Operations, 38
   Maintenance and Spare Parts, 38
   General Management Issues, 39
   Programmatic Issues, 39
  Pre-Operational Survey, 40
  Disposal Program Staffing, 43

4 SYSTEMIZATION ENVIRONMENTAL PERFORMANCE 45
  TOCDF Permitting Requirements, 45
  Surrogate Trial Burns, 45
   Liquid Incinerator #1, 45
   Deactivation Furnace System Surrogate Trial Burn, 46

5 COMMUNITY INTERACTION AND PLANNING. 48
  Utah Community Involvement, 48
   Utah Citizens Advisory Commission and Risk Assessment: Problems of
    Communication, 49
   Personal Protective Equipment, 50
  Community Emergency Planning, 51
   Training, 51
   Emergency Planning, 52
   Emergency Communications, 52
   Emergency Medical Care, 53
  Army Citizens Involvement Program in Utah, 53

6 OVERVIEW OF SITE-SPECIFIC RISK ASSESSMENT 55
  NRC Recommendations on Risk Management, 55
   Report on Operational Verification Testing, 56
   NRC Letter Report on the Chemical Stockpile Disposal Program Risk
    Management Process, 56
   Recommendations Report, 58
  TOCDF Risk Management Plan, 59
   Operation of Metal Parts Furnace Feed Airlock, 61
   Weteye Bomb Aluminum and Agent Interaction, 62
   Weteye Bomb Handling and Inventories, 62
   Seismic Anchorage of the Liquid Propane Gas Tank, 62
  Tooele Risk Assessment, 62
  Accident Quantitative Risk Assessment, 63
   Methodology, 63
   Independent Review Committee Role and Evaluation, 65
   Results, 66
  NRC Evaluation of the Accident Risk Program, 69

7 FINDINGS AND RECOMMENDATIONS 70
   Overview, 70
   Findings, 70
      Responses to OVT II Report Recommendations, 70
      Responses to Monitoring Report Recommendations, 72
      Responses to Risk Letter Report Recommendations, 74
      Responses to the Recommendations Report, 76
   Recommendations, 78
      Duration of TOCDF Operations, 78
      Coordinated with the Start of Agent Operations, 78
      Prior to the Start of Agent Operations, 79
      During the First Year of Agent Operations, 79

APPENDICES
   A Public Law 102-484—Oct. 23, 1992 (Extract), 83
   B Chemical Stockpile Disposal Program, 84
      The Call for Disposal, 84
      Disposal Program Background and Role of the National Research Council, 84
      Description of the Stockpile, 85
         Agents, 85
         Containers and Munitions, 86
         Geographical Distribution, 87
      The Baseline Incineration System, 87
         Storage, Transportation, and Unloading of Munitions and Containers, 87
         Disassembly and Draining, 88
         Agent Destruction, 89
         Destruction of Energetics, 90
         Metal Parts Decontamination, 91
         Pollution Abatement Systems, 93
         Auxiliary Systems, 94
         Agent Monitoring Systems, 95
   C Recommendations of the Committee on Review and Evaluation of the Army
        Chemical Stockpile Disposal Program (Stockpile Committee), 96
   D Public Meeting, Tooele County Courthouse, Tooele, Utah, 105
      Agenda, 106
      Letters of Invitation, 108
         Citizens Advisory Commission Invitation Letter, 108
         Public Invitation Letter, 109
      Distribution List, 110
   E Biographical Sketches, 113

REFERENCES 117

Tables and Figures

TABLES

3-1   Severity Ranking Criteria, 29
3-2   Criteria Used to Establish Qualitative Frequency Categories, 30
3-3   Risk Assessment Code (RAC), 31
3-4   TODCF Pre-Operational Survey Team Members, 41
4-1   Summary of Results from the TOCDF Liquid Incinerator #1 Surrogate Trial Burn, 46
6-1   Reports Associated with the Expert Panel Review of the Tooele Chemical Agent Disposal Facility Quantitative Risk Assessment, 67
6-2   Presentations to the Expert Panel Review of the Tooele Chemical Agent Disposal Facility Quantitative Risk Assessment, 68
B-1   Composition of Munitions in the U.S. Chemical Stockpile, 88
B-2   Chemical Munitions Stored in the Continental United States, 90
B-3   Approximate Amounts of Metals, Energetics, and Agent Contained in the Unitary Chemical Stockpile (tons), by Site, 91
B-4   Air and Exposure Standards, 94
C-1   Recommendations from *Evaluation of the Johnston Atoll Chemical Agent Disposal System Operational Verification Testing: Part I* (OVT 1) and *Part II* (OVT 2), 96
C-2   Recommendations from *Review of Monitoring Activities Within the Army Chemical Stockpile Disposal Program* (MON), 98
C-3   Recommendations from the Letter Report to the Assistant Secretary of the Army to Recommend Specific Actions to Further Enhance the CSDP Risk Management Process (RISK), 99
C-4   Recommendations (REC) and Findings (FIND) from *Recommendations for the Disposal of Chemical Agents and Munitions*, 100

FIGURES

1-1   Schematic drawing of the baseline incineration system, 9
3-1   Outline of the Facility Construction Certification Process, 33
6-1   Overview of the Risk Management Plan, 60
6-2   Hierarchy of regulations that define safety at the TOCDF, 61
6-3   Identifying upsets, 64
6-4   Sample portion of a rocket handling process operational diagram, 65
6-5   Schematic drawing of process operational diagram development, 66
B-1   M55 rocket and M23 land mine, 86
B-2   105-mm, 155-mm, 8-inch, and 4.2-inch projectiles, 87
B-3   Bomb, spray tank, and ton container, 87
B-4   Types of agent and munitions and percentage of total agent stockpile (by weight of agent) at each storage site, 89
B-5   Schematic drawing of the baseline system, 92
B-6   Schematic drawing of a pollution abatement system, 93

Abbreviations and Acronyms

ACAMS	Automatic Continuous Air Monitoring System
ACS	Agent Collection System
AED	Atomic Emission Detector
AQS	Agent Quantification System
BDS	Bulk Drain Station
BRA	Brine Reduction Area
CAC	Citizens Advisory Commission
CAMDS	Chemical Agent Munitions Disposal System, Tooele, Utah
CDTF	Chemical Demilitarization Training Facility, Aberdeen, Maryland
CEM	Comprehensive Emergency Management (Utah Division of)
CEMS	Continuous Emission Monitoring System
CFR	Code of Federal Regulations
CHB	Container Handling Building
CPRP	Chemical Personnel Reliability Program
CSDP	Chemical Stockpile Disposal Program
CSEPP	Chemical Stockpile Emergency Preparedness Program
DAAMS	Depot Area Air Monitoring System
DEQ	Department of Environmental Quality
DFS	Deactivation Furnace System
DoD	Department of Defense
DPE	Demilitarization Protective Ensemble
DRE	Destruction Removal Efficiency
dscm	Dry standard cubic meter
DSHW	Division of Solid and Hazardous Waste (Utah)
DUN	Dunnage Furnace
EG&G	Edgerton, Germerhausen and Grier, Inc.
ENVCP	Environmental Compliance Plan
EOC	Emergency Operations Center
EPA	Environmental Protection Agency
EPZ	Emergency Planning Zone
ETA	Event Tree Analysis
FEMA	Federal Emergency Management Agency
FLLRT	Field Lessons Learned Review Team
FMEA	Failure Modes and Effects Analysis
FPEIS	Final Programmatic Environmental Impact Study
FTA	Fault Tree Analysis
FTIR	Fourier Transform Infrared

GA	Tabun
GB	Sarin
GC/FPD	Gas Chromatograph with Flame Photometric Detector
GC/MSD	Gas Chromatograph with Mass Spectrometric Detector
H, HD, HT	Blister or Mustard Agents
HEPA	High-Efficiency Particulate Air
HVAC	Heating, Ventilating, and Air Conditioning
HWMU	Hazardous Waste Management Unit
ID	Induced Draft
IDLH	Immediately Dangerous to Life and Health
in.	inch
IRZ	Immediate Response Zone
JACADS	Johnston Atoll Chemical Agent Disposal System
lb	pound
LIC	Liquid Incinerator
m^3	cubic meter
mg	milligram
µg	microgram
MHz	Megahertz
min	minute
mm	millimeter
mM	millimolar
MPF	Metal Parts Furnace
ng/kg	nanogram per kilogram
NO_x	Nitrogen Oxides
NRC	National Research Council
OSHA	Occupational Safety and Health Administration
OTA	Office of Technology Assessment
OVT	Operational Verification Testing
PAS	Pollution Abatement System
PCB	Polychlorinated Biphenyl
PCDD/F	Polychlorinated Dibenzo-p-dioxins and Dibenzofurans
PIC	Product of Incomplete Combustion
PLL	Programmatic Lessons Learned
PMCD	Program Manager for Chemical Demilitarization
PM-CSD	Project Manager for Chemical Stockpile Disposal
PMD	Projectile/Mortar Disassembly Machine
POD	Process Operational Diagram
POHC	Principal Organic Hazardous Constituent
PPE	Personal Protective Equipment
ppm	parts per million

QRA	Quantitative Risk Assessment
RAC	Risk Assessment Code
RCRA	Resource Conservation and Recovery Act
RSM	Rocket Shear Machine
RMP	Risk Management Plan
s	second
SAIC	Science Applications International Corporation
SAR	Subject Area Review
SDS	Spent Decontamination System
SHA	Systems Hazard Analysis
SO_x	Sulfur Oxides
TOCDF	Tooele Chemical Agent Disposal Facility
TSCA	Toxic Substances Control Act
TWA	Time-Weighted Average
USACDRA	U.S. Army Chemical Demilitarization and Remediation Activity
USACMDA	U.S. Army Chemical Materiel Destruction Agency
USATHAMA	U.S. Army Toxic and Hazardous Materials Agency
VX	Organophosphate Nerve Agent
3X	Three-X Level of Decontamination
5X	Five-X Level of Decontamination

Executive Summary

The Army Chemical Stockpile Disposal Program and earlier chemical demilitarization activities have had a long history. National Research Council (NRC) committees have been involved since 1969. The Army's development of its baseline incineration system for the destruction of the nation's stockpile of lethal unitary chemical agents and munitions has its roots in research and development activities that took place at least as early as the late 1970s and early 1980s.[1] More recently, in 1992, Congress, in Public Law 102-484 (appendix A), directed the Army to dispose of the entire unitary chemical warfare agent and munitions stockpile by the end of 2004. Appendix B contains a brief history of the Chemical Stockpile Disposal Program (CSDP), of NRC reporting on that program, and a description of the principal elements and components of the baseline system.

The two extant examples of the baseline incineration system are the Johnston Atoll Chemical Agent Disposal System (JACADS), located on Johnston Island in the Pacific Ocean more than 700 miles southwest of Hawaii, and the Tooele Chemical Agent Disposal Facility (TOCDF) located at the Tooele Army Depot in Utah about 80 miles west of Salt Lake City. JACADS was established as the first full-scale version of the baseline system and commenced Operational Verification Testing in July 1990. Construction of the TOCDF, the first baseline system disposal facility located in the continental United States, was begun in 1989, and systemization was begun in August 1993.

At the request of the Secretary of the Army, the NRC Committee on Review and Evaluation of the Army Chemical Stockpile Disposal Program (Stockpile Committee) was established in 1987 to provide the Army with technical advice and counsel on the Chemical Stockpile Disposal Program. In evaluating the adequacy of testing and operations at JACADS, the committee produced the following reports in 1993 and 1994 (whose recommendations are summarized in appendix C):

- *Evaluation of the Johnston Atoll Chemical Agent Disposal System Operational Verification Testing: Part I and Part II;*
- *Review of Monitoring Activities Within the Army Chemical Stockpile Disposal Program;*
- Letter report to the Assistant Secretary of the Army to recommend specific actions to further enhance the CSDP risk management process; and
- *Recommendations for the Disposal of Chemical Agents and Munitions.*

Based on the recommendations contained in the above reports, this report assesses the Army's changes and improvements to the safety and operations of the baseline incineration system and the associated monitoring system as implemented during systemization of the TOCDF. In addition, this report reviews the conduct of the Army site-specific risk assessment for the TOCDF. The recommendations in this report indicate actions the Army should take both prior to the start of and during the first year of agent operations at the TOCDF.

At this writing, the start of agent operations at the TOCDF is planned for the first quarter of 1996. Although the Army has the overall responsibility for the decision whether to start agent operations (subject to meeting all regulatory requirements), the Stockpile Committee was requested to provide this report as additional input to that decision. Specifically, the report has been prepared in response to the Army's informal

---

[1]The term unitary distinguishes a single chemical loaded in munitions or stored as a lethal material. More recently, binary munitions have been produced in which two relatively safe chemicals are loaded in separate compartments to be mixed to form a lethal agent after the munition is fired or released. The components of binary munitions are stockpiled apart, in separate states. They are not included in the present Chemical Stockpile Disposal Program. However, under the Chemical Weapons Convention of 1993, they are included in the munitions that will be destroyed.

request to the NRC for a review and assessment of the systemization of the TOCDF.

In developing the findings and recommendations in this report, the Stockpile Committee found it necessary to evaluate the safety and environmental performance of the TOCDF systemization. Safety performance was assessed by direct committee observation and the evaluation of third party information (e.g., published reports prepared by the Army, Army contractors, and state agencies as well as discussions with plant personnel, citizens, and regulators) obtained by the committee. In addition, the committee assessed the Army's Pre-Operational Survey, which was to be used as a basis for making the decision to start agent operations at the TOCDF. Environmental performance was assessed primarily by evaluating the results of surrogate trial burns demonstrating that each component of the baseline incineration system destroys selected compounds to a destruction removal efficiency of 99.9999 percent (6-nines).

The Stockpile Committee continues to monitor the Army's public affairs and community relations program for the Chemical Stockpile Disposal Program. The committee offers observations on the Army's efforts to encourage local community involvement and to work closely with the Utah Citizens Advisory Commission and responsible emergency management agencies on the Chemical Stockpile Emergency Preparedness Program (CSEPP) in the state of Utah.

Finally, the committee offers observations on the Army's response to the committee's risk-related recommendations. The committee describes the Army's plans for risk assessment and risk management at the TOCDF as well as the current status of those activities.

FINDINGS

The Stockpile Committee has completed this review of the present status of the TOCDF design modifications and systemization. This review is based on the committee's knowledge of the baseline system, on information provided by the Army and others, and on four site visits to the TOCDF, located at the Tooele Army Depot, Utah. The visits took place in November 1991, March 1993, May 1994 (shortly after the start of systemization), and March 1995 (towards the end of systemization). In addition, four subgroups of the committee made separate visits to the site between February and June 1995. The findings and recommendations presented here are based on information obtained prior to September 30, 1995, before the completion of all requirements that must be met prior to the start of agent operations. The recommendations are based on the committee's knowledge of the current status of the TOCDF and on the committee's understanding of the actions that must be completed before the Army authorizes the start of agent operations. As of this writing, some final activities still require completion by the Army and Army contractors.

Finding 1. The Stockpile Committee finds that the Army has implemented or will soon implement the changes recommended in *Evaluation of the Johnston Atoll Chemical Agent Disposal System Operational Verification Testing: Part II*.

Except for reducing the confirmation time for false positive alarms and establishing a system to track laboratory errors, the recommended monitoring system improvements have been made. The reduction of confirmation time is a difficult problem, but efforts are under way to solve it. The system for laboratory operations error analysis was scheduled for implementation by December 1995.

The brine reduction area and the dunnage furnace have been modified and have passed systemization tests at the TOCDF; they will be tested and fully certified shortly after the start of agent operations. The TOCDF liquid incinerator is expected to be capable of meeting nitrogen oxides emission requirements without an additional abatement system.

A hot-slag removal system has been installed on each liquid incinerator. These have not been tested under hot-slag conditions because the Army has decided that there is no adequate surrogate test material for the highly variable slag to allow for complete testing of the system. The Army will test the system as slag accumulates from agent-destruction operations. If the slag removal system does not allow for hot-slag removal, the current manual removal procedure will be resumed.

MITRE Corporation has performed a detailed study (MITRE, 1994) and concluded that changes made to the munitions handling system minimize the chances of a recurrence of the misfeeding problems that occurred at JACADS. The current system is also being analyzed in detail in the risk assessment study being performed by Science Applications International Corporation (SAIC), providing an independent review of the safety of the munitions handling systems.

The Army has developed an improved agent-extraction verification system to handle gelled mustard and to avoid the need for manual intervention in routine operations.

The TOCDF has organized a program management unit, staffed by the Army, to focus on safety, quality assurance, and environmental oversight. A similar organization exists within the Edgerton, Germerhausen and Grier, Inc. (EG&G) program management organization (the prime contractor for the TOCDF). Environmental oversight includes both permitting activities and environmental compliance. SAIC has been retained by the Army to provide additional staff through an environmental compliance field office. Close working relationships have been established with the Utah Department of Environmental Quality to maintain responsive interactions to facilitate preparation of the Resource Conservation and Recovery Act (RCRA) permit as the TOCDF makes final modifications in response to the systemization experience and other factors. These organizations are charged to maintain compliance with all environmental regulations throughout the lifetime of the facility.

The management structures in place at the TOCDF and limited evidence gathered by Stockpile Committee members during site visits suggest that high-quality safety management systems for agent operations are in place, although the committee believes that more attention must be paid to nonagent safety issues to promote a total safety culture at the facility. All action items identified as incomplete in the Pre-Operational Survey must be resolved by the Army prior to the start of agent operations. The Army has initiated a Programmatic Lessons Learned program, created a Field Lessons Learned Review Team, commenced Subject Area Reviews, and established a Risk Management Plan which is being developed by SAIC as an outgrowth of their TOCDF risk assessment work. The Army has developed a process to identify "precursor events" at operating facilities (e.g., equipment failures and human errors) and suggest improvements to maintain a continuing emphasis on a philosophy of totally safe operations.

The TOCDF risk assessment has been performed by SAIC under the observation of an outside Expert Panel. One Stockpile Committee member monitored most of the Expert Panel meetings. Because of time constraints, the entire risk assessment will not be completed before the start of agent operations at the TOCDF. However, risk assessment pertinent to each succeeding campaign (i.e., destruction of a particular munition or container type and a particular agent) will be completed and lessons learned will be applied to the facility before the start of each campaign. The first analysis for campaigns 1 and 2 (disposal of GB and VX M55 rockets with co-processing of bulk items) was completed in April 1995. A few resulting modifications to the TOCDF and its operations are being made and will be completed before the start of agent operations. The modified final analysis for the first two campaigns was completed on June 26, 1995. The complete TOCDF risk assessment is expected to be published in the first quarter of 1996. The committee believes that the Army's approach to the TOCDF risk assessment for the first two campaigns meets the spirit of the committee's recommendation.

Finding 2. The Stockpile Committee finds that the Army has implemented or will soon implement most of the changes recommended in *Review of Monitoring Activities Within the Army Chemical Stockpile Disposal Program.*

Capability for positive identification of chemical agent species has been added to the laboratory with the new gas chromatograph/mass spectrometer. This capability is not presently available with field monitors.

The need for continuous monitoring capability for all agents has been met by placing Automatic Continuous Air Monitoring System (ACAMS) monitors calibrated for the various agents in the unpack areas of the TOCDF. A multiagent ACAMS is under development. Continuous monitoring is not yet in place in storage areas, but the Army has told the Stockpile Committee that plans are under way to implement continuous monitoring there.

The need to reduce the time for confirmation of false positives has not been met. ACAMS alarms still require the laboratory analysis of samples from the Depot Area Air Monitoring System (DAAMS) to confirm a false positive. A single unconfirmed alarm requires shutdown of agent operations but does not by itself initiate the response appropriate for a major agent release. False positive signals result in plant disruptions and increase the potential for human error and equipment degradation. The Army expects that the multiagent ACAMS will have a lower false positive rate. (A dual detector ACAMS is expected to be ready for field tests in December 1995).

Procedures for periodic testing of field sensors to eliminate false negatives if a significant release should occur are based on a more comprehensive schedule for testing field sensors. Some monitors have also been installed that can reliably detect higher levels of agent should a significant release occur.

Faster response monitors, set to higher detection levels (immediately dangerous to life and health), have been installed in the unpack area of the facility. A new type of detector (based on ion mobility) with about a 30-second response time is presently being developed.

The TOCDF will undergo a series of required RCRA and Toxic Substances Control Act state-supervised trial burns for the first 24 months of operations; a minimum of 16 trial burns (each consisting of three or four runs) are scheduled. The results of the trial burns will be used to revise the Health Risk Assessment and to formulate a monitoring approach for products of incomplete combustion, particulates, heavy metals, halogenated dioxins and furans, and all volatile and semivolatile organics.

The TOCDF laboratory uses a new bar code system for identifying samples and an improved automated record system. The Stockpile Committee recognizes major improvements in quality control over the procedures at JACADS but reiterates the need for an additional system to track and evaluate lab errors. The additional system would also provide input for continuing improvements to laboratory operations. Such a system is scheduled to be in place by December 1995.

Laboratory personnel are now given a variety of tasks. In addition to analyzing for agents, they now perform nonagent analysis of flue gas samples using mass spectrometry and gas chromatography. Laboratory operators are not presently cross-trained in all of these techniques. However, in the next few months the Army plans to start cross-training in the operation of various instruments, handling hazardous waste, testing new monitors, and performing statistical analyses of data. Cross-training will enhance personnel attentiveness and enhance individual job performance.

The Army has established and implemented a system for providing double blind challenges to the laboratory in a way that will provide frequent and valid checks on the quality and reliability of normal laboratory operations.

Finding 3. The Stockpile Committee finds that the Army has implemented or will soon implement the analyses and actions recommended in the risk letter report.

SAIC completed the first phase of an accident quantitative risk assessment for the TOCDF, and a summary of the results was provided to the committee on June 26, 1995. This first phase covers the portions of the facility that will be involved in the first two campaigns at the facility (disposal of GB and VX M55 rockets with co-processing of bulk items). The full accident quantitative risk assessment for storage risks and remaining campaigns is expected to be completed in the first quarter of 1996. The techniques used and scenarios considered in the quantitative risk assessment are in accordance with the Stockpile Committee's recommendations. Latent health effects from nonagent exposures are not included in SAIC's scope of work. The State of Utah is analyzing latent health effects independently, following Environmental Protection Agency guidelines for Health Risk Assessments.

SAIC has recognized capability for performing risk assessments but is involved in the Chemical Stockpile Disposal Program because it has been a long-term support contractor for the Army. The Army reports that SAIC was chosen to avoid a long procurement process to select an independent contractor, which would have introduced delays in the program. The Army retained another long-term support contractor, MITRE Corporation, to organize an Expert Panel to oversee the SAIC risk assessment. Five experts were chosen for the panel. The Stockpile Committee has reviewed their qualifications, and they are all well qualified. One of the members is a combustion expert from Brigham Young University in Salt Lake City, who provides some degree of local perspective for the panel. At the request of the Army, the Stockpile Committee was invited to have a member who is an expert in risk assessment monitor the meetings of the Expert Panel. The Stockpile Committee believes that the Army has complied with the intent of the recommendation that an independent risk assessment be made.

Although SAIC briefed the Utah Citizens Advisory Commission in July 1994 on its methodology and plans for conducting the risk assessment, they have not established a dialogue that is perceived by the committee to be interactive with either that group or with other interested local parties. Additional information on local concerns about the quantitative risk assessment was obtained from newspaper articles. The committee believes interactive communication between the risk assessment contractor and local community groups about future risk analyses is essential.

Several risks were identified as a result of the detailed risk assessment, and appropriate changes in equipment and procedures were made to mitigate them.

Finding 4. The Stockpile Committee finds that the Army has implemented or will soon implement changes pertinent to the TOCDF that were recommended in *Recommendations for the Disposal of Chemical Agents and Munitions*.

The *Recommendations* report reiterated the recommendations from the *OVT, Monitoring*, and site-specific risk assessment reports. Three of the recommendations pertaining to the TOCDF were related to public involvement, carbon filtration, and program staffing.

Public Involvement

The committee finds that the Army's efforts in Utah to obtain community input into the risk assessments were substantial, but not especially productive. The committee believes that such involvement prior to the beginning of a risk assessment, as well as during its implementation, is essential to improving risk communication and gaining public acceptance of the results.

In March 1995, the committee found that the long-delayed approval of personal protective equipment for emergency management responders, associated training, and funding for core response personnel continued to be roadblocks to implementation of the Chemical Stockpile Emergency Preparedness Program (CSEPP) by both the Utah Division of Comprehensive Emergency Management and the Tooele County Department of Emergency Management. At this writing, the personal protective equipment has been approved, but unresolved issues remain pertaining to training in the use of the equipment.

The risks presented by the stockpile require that emergency response plans be completed and exercised to ensure preparedness and successful response should there be an actual release. The Utah Division of Comprehensive Emergency Management has indicated that the absence of national planning standards has meant insufficient guidelines for reentry, emergency medical services, and recovery phase operations, and that this in turn has led to a lack of fully effective training and exercises. Moreover, some counties opted not to participate in exercises, raising concerns about their level of preparedness for an emergency. The lack of national planning standards cannot be permitted to interfere with planning for public safety and, if necessary, the Army must step in to rectify the situation.

The committee finds that the local CSEPP emergency planning efforts are incomplete as is evident by the fact that the Tooele County Emergency Operations Plan appendices are still in draft form.

The committee also finds that the Communication Plan for Tooele County and the planned implementation of the communications system linking the important operations centers in the emergency planning zone are not yet complete. In addition, as of this writing, public notification tone alert radios are not yet in place.

The Army has begun to implement a large and comprehensive public information program in Tooele. This program is impressive because few activities and resources had been devoted to public information and outreach until recently. Yet the list of activities, either planned or under way, still suggests that not enough attention is being paid to soliciting citizen input for programmatic decisions. Additionally, public concerns about emergency management issues are likely to increase when the TOCDF starts agent operations. Efforts should be made by the Army to include the public in discussions of CSEPP issues and to integrate better the public outreach program with elements of the CSEPP program.

Carbon Filtration

The Army is currently evaluating whether the installation of carbon filter equipment for incineration exhaust streams at continental U.S. disposal sites would be warranted. If so, the Army has chosen the TOCDF as the site for a demonstration unit, which would be located on one of the TOCDF incinerator exhaust gas systems. Space for the addition of systems to all incinerator exhaust gas streams has been kept available at the TOCDF in the event that carbon filters are warranted for the TOCDF site-specific conditions. Studies are planned to evaluate whether the installation of carbon filtration systems would be warranted at any of the other continental U.S. sites after the TOCDF evaluation program is completed.

Program Staffing

The committee has observed the addition of qualified personnel, both in the office of the Program Manager for Chemical Demilitarization (PMCD) and in contractor organizations at the TOCDF. The present level of staffing at the TOCDF appears appropriate for safe and environmentally compliant operation of the facility. Additional PMCD personnel may be needed as operations expand.

The retirement of the Tooele Chemical Agent Disposal Facility EG&G general manager so close to the start of agent operations raised committee concerns. A new general manager has since been appointed, and he worked with his predecessor during the months of July and August 1995 to facilitate the transition. The retiring general manager will remain on call as a consultant through the start of agent operations.

RECOMMENDATIONS

Based on the Stockpile Committee's evaluation of the status of the TOCDF with respect to recommendations made in previous reports, the committee is generally satisfied with the progress made and recommends the following actions pertaining to safety and performance be taken at the TOCDF:

Duration of TOCDF Operations

Recommendation 1. The development and implementation of the overall safety program at the TOCDF must be given high priority.

Recommendation 2. Safety and environmental performance goals should be given at least equal weight with production goals in establishing award fee criteria.

Recommendation 3. Applicable portions of the accident quantitative risk assessments must be completed and all safety-related concerns resolved before the start of specific agent-destruction campaigns.

Recommendation 4. A substantial effort should be made by the Army to enhance interactive communications with the host community and the Utah State Citizens Advisory Commission on issues of mutual concern (e.g., various elements of the Chemical Stockpile Emergency Preparedness Program (CSEPP), decontamination and decommissioning, future use of the facility, and risk reduction).

Coordinated with the Start of Agent Operations

Recommendation 5. The Army should increase efforts to work with the Utah Division of Comprehensive Emergency Management to ensure that first-responders have been adequately trained to use the personal protective equipment approved by the Occupational Safety and Health Administration. Tooele County must ensure their capability for responding to an emergency, especially because this condition relates to state requirements for the start of agent operations.

Recommendation 6. The Army, and where appropriate the Federal Emergency Management Agency (FEMA), should ensure that local and state Chemical Stockpile Emergency Preparedness Program plans for responding to potential chemical events are complete and well exercised as soon as possible.

Recommendation 7. The Army/FEMA should provide the necessary resources for completing the communications system planned by the Tooele County Department of Emergency Management.

Prior to the Start of Agent Operations

Recommendation 8. All mandatory requirements of the Army's Pre-Operational Survey must be satisfied.

Recommendation 9. The liquid incinerator and deactivation furnace system must have demonstrated a destruction removal efficiency of 99.9999 percent (6-nines) during surrogate trial burns.

Recommendation 10. High quality, adequately staffed safety management systems must be completely implemented (including procedures for testing critical equipment; all necessary operating, maintenance, and emergency procedures; management of change procedures; training and cross-training programs; programmatic lessons learned activities; subject area reviews; and other safety oversight activities).

During the First Year of Agent Operations

Recommendation 11. The liquid incinerator and the deactivation furnace system must pass all required Resource Conservation and

Recovery Act trial burns; the deactivation furnace system must also pass required Toxic Substances Control Act trial burns.

Recommendation 12. Testing and certification of the brine reduction area and the dunnage incinerator should be completed at the TOCDF, or a satisfactory disposal alternative must be implemented.

Recommendation 13. Performance of the slag removal system for the liquid incinerators should be demonstrated when sufficient slag has accumulated.

Recommendation 14. The Risk Management Plan must be fully implemented.

Recommendation 15. A comprehensive, integrated, and clear TOCDF risk assessment study, including a full description of all significant acute and latent agent and nonagent risks associated with disposal operations, as well as with the continued maintenance of the Tooele chemical stockpile, should be completed. A full explanation of the uncertainties associated with the various estimates should be included.

Recommendation 16. A system for documenting and tracking unexpected upsets, errors, failures, and other sources of problems that have led to "near misses" during operation of the facility should be developed as soon as possible. A program for integrating this information into a plan for continual safety improvements at the TOCDF should be implemented.

Recommendation 17. An active program for continual improvement of monitoring instrumentation, including techniques for more rapid recognition of significant levels of agent release, should be pursued.

Recommendation 18. Evaluations of the stack-gas carbon filter bed system should be continued.

1

Introduction

This report is in response to an Army request for a review and assessment of the systemization of the Tooele Chemical Agent Disposal Facility (TOCDF), located at Tooele Army Depot, Utah. The report emphasizes issues raised in earlier reports by the National Research Council (NRC) Committee on Review and Evaluation of the Army Chemical Stockpile Disposal Program (Stockpile Committee) in 1993 and 1994.

Start of agent operations at the TOCDF is now scheduled for the first quarter of 1996. Although the Army has the overall responsibility for the decision to start agent operations (subject to meeting all regulatory requirements), the Stockpile Committee was asked to provide this report as further input to that decision. Specifically, the Army requested that the Stockpile Committee assess, prior to the start of agent operations, the Army's implementation of the committee's recommended changes and improvements in the safety and operations of the TOCDF.

To complete this report prior to the start of agent operations, the Stockpile Committee found it necessary to cut off the search for data and information after September 30, 1995. As of that date, some activities by the Army and Army contractors were still incomplete. However, procedures are already in place to assure and verify completion of the necessary tasks. Consequently, this report is based on the committee's review of the status of the facility several months before the scheduled start of agent operations and on the committee's understanding of the requirements that will be imposed by the Army and regulatory authorities before agent operations are authorized by the Army.

Chapter 2 describes the improvements implemented in the TOCDF design and operations as of the date of this report. Chapters 3 and 4 present the Stockpile Committee's evaluation of the safety and environmental performance during the TOCDF systemization. Chapter 5 discusses public involvement and concerns with respect to the TOCDF, and chapter 6 addresses the new risk assessment studies for the TOCDF. Chapter 7 presents findings and recommendations.

CHEMICAL STOCKPILE DISPOSAL PROGRAM

Since 1969, the Army and various committees of the National Research Council have addressed problems associated with eliminating the nation's stockpile of lethal unitary chemical agents and munitions. In 1985, the Army established the Chemical Stockpile Disposal Program (CSDP). In 1987, the present Stockpile Committee was established by the NRC at the request of the Secretary of the Army. The NRC efforts have produced a series of reports evaluating many aspects of the CSDP. A brief history of this mutual effort can be found in appendix B.

The Unitary Chemical Agent and Munitions Stockpile

There are two basic types of chemical agents in the unitary stockpile: nerve agents (GB, VX) and blister agents (mustard). The agents are contained in a variety of bulk containers and munitions. Munitions such as M55 rockets and various other projectiles have associated explosives and propellants (so-called "energetics"), which also require disposal. The amount and types of agents, energetic materials, and associated metal containers differ from site to site. The stockpile at the Tooele Army Depot includes all types of agent, munitions, and bulk containers.

Fundamentals of Disposal

Disposing of chemical agents and munitions means releasing all unitary stockpile materials from Army control after they have been altered to satisfy both international treaty requirements and domestic environmental requirements. Waste streams from disposal processes may be gaseous, liquid, or solid. A number of federal and state environmental regulations govern disposal operations in the continental United States, including the Resource Conservation and Recovery Act (RCRA), the Toxic Substances Control Act (TSCA), the

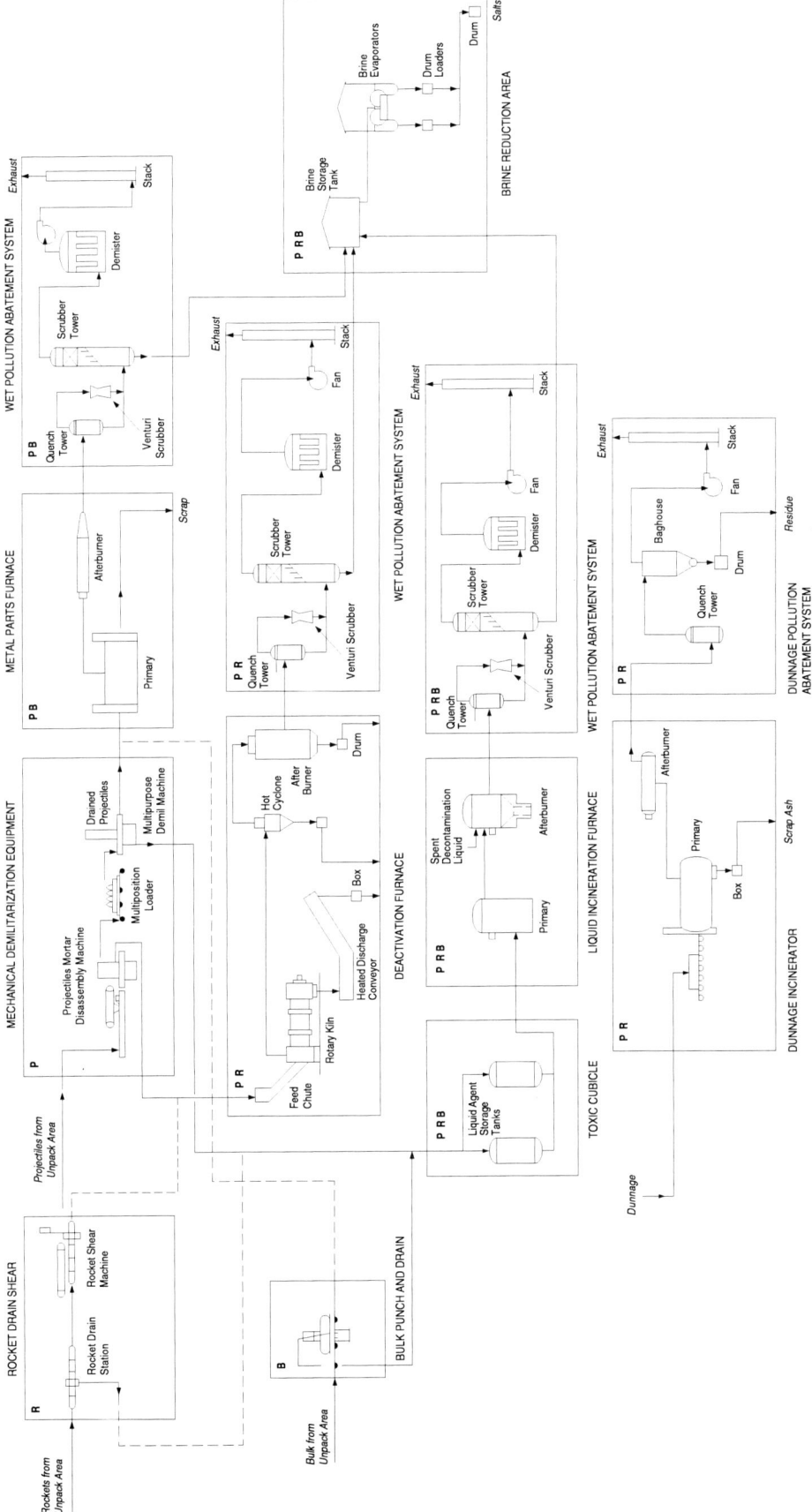

FIGURE 1-1 Schematic drawing of the baseline incineration system (U.S. Army, 1988; NRC, 1994a, 1994b, 1994c).

Clean Air Act, the Clean Water Act, and the Hazardous Materials Transportation Act.

The Baseline Incineration System

The performance and safety of candidate disposal methods are greatly increased if stockpile feed materials are separated into distinct streams of agent, energetic materials, metal parts, and dunnage (packing and other miscellaneous material) prior to disposal or destruction. A schematic drawing of the baseline incineration system is shown in figure 1-1.

In the baseline system, these materials are separated inside a building that has areas capable of withstanding explosions. The pressure in these and other areas is kept below atmospheric pressure to prevent agents from leaking to the atmosphere. Agents are removed from munitions and containers by two methods. Most containers are simply mechanically punched open and drained, whereas artillery projectiles are mechanically disassembled and drained. These processes yield three material streams: agent, energetics, and metal parts. Energetics and metal parts may be contaminated by agents, although the vast majority (95 percent or more) of the agent is recovered separately. This material separation is a major safety feature of the baseline system, which allows for the design and operation of parallel disposal components that are properly matched to the three widely differing material streams.

Agents are pumped to and destroyed in a specially designed liquid incinerator (LIC). This incinerator consists of primary and secondary combustion chambers and a pollution abatement system (PAS). Agent flow is stopped if the combustion chamber temperature drops below 2,550°F. Associated energetics are burned in a rotary kiln deactivation furnace system; exhaust gases are sent to an afterburner and then treated by a pollution abatement system. Metal parts are decontaminated by heating them in a metal parts furnace (MPF) to 1,000°F for a minimum of 15 minutes. Any residual agent is vaporized from the metal parts and burned within the furnace. Exhaust gases are sent to an afterburner and then to a pollution abatement system that removes gaseous pollutants and particulates.

Agent compounds contain various inorganic atoms that result in significant acid gas products. Acidic gases are scrubbed from the discharge stream with alkali solutions to form salts. These wet salts, or "brine," are dried in a brine reduction area (BRA), and the resultant dry salts are stored for later disposal in hazardous waste landfills.

Contaminated and noncontaminated packing materials and miscellaneous waste, or "dunnage," are burned in a dunnage furnace (DUN). Because only trace amounts of agent may be present, dunnage furnace combustion produces low acid gas concentrations that do not require treatment prior to release. Exhaust gases are discharged through a separate stack without acid gas scrubbing.

Two auxiliary material streams are also processed: decontamination fluids are processed in the secondary combustion chamber of the liquid incinerator; and ventilation air is passed through banks of charcoal filters to remove any trace contaminants.

The baseline monitoring system is used to detect the release of agents and to monitor adherence to all environmental requirements. The system consists of a combination of the Automatic Continuous Air Monitoring System (ACAMS), which detects immediate threats (three to eight minute response time at 20 percent of permissible eight-hour exposure level for workers, see table B-4, appendix B), and the Depot Area Air Monitoring System (DAAMS), which has a slower response time but provides a much more sensitive measurement. ACAMS alarms require immediate agent-feed shutoff. DAAMS analyses are used to confirm ACAMS alarms and to document environmental compliance.

A comprehensive description of the baseline incineration system is provided in appendix B.

SUMMARY

Additional information on the chemical agents in the U.S. stockpile as well as on associated munitions and containers is presented in appendix B, which outlines inventories at the Tooele Army Depot and presents an overview of the baseline incineration system as implemented and tested at Johnston Atoll Chemical Agent Disposal System (JACADS). Comprehensive data on the stockpile can also be found in a number of other reports (NRC, 1984; U.S. Army, 1988). The baseline system is described in more detail in several reports (MITRE, 1991, 1992, 1993a, 1993b, 1993c; NRC, 1994a).

Pertinent findings and recommendations from prior Stockpile Committee reports can be found in appendix C. To assist the reader, an alpha-numeric code has been assigned to each of these findings and recommendations and accompanies their use throughout this report.

2

Changes at the Tooele Chemical Agent Disposal Facility in Response to National Research Council Recommendations

REVIEW OF NRC RECOMMENDATIONS REGARDING OPERATIONAL VERIFICATION TESTING AT JACADS

In 1994, the National Research Council (NRC) Committee on Review and Evaluation of the Army Chemical Stockpile Disposal Program (Stockpile Committee) published its report *Evaluation of the Johnston Atoll Chemical Agent Disposal System Operational Verification Testing: Part II* (NRC, 1994a). This report contained a general recommendation to proceed with systemization of the Tooele Chemical Agent Disposal Facility (TOCDF) and, during systemization, to conduct needed tests and make improvements (this recommendation is coded OVT2-2 in the present report; see appendix C). The specific tests and improvements listed in this recommendation are reviewed in the following sections of this chapter.

Brine Reduction Area

Satisfactory operation of the brine reduction area was not demonstrated during the initial JACADS Operational Verification Testing (OVT). As a result, the Stockpile Committee recommended that this activity be addressed satisfactorily prior to the start-up of the TOCDF:

Complete the brine reduction area (to include its pollution abatement system) performance tests, or develop a satisfactory brine disposal alternative. (OVT2-2B[1])

The Army subsequently made a number of equipment modifications and completed the JACADS test. The Army report (U.S. Army, 1995a) shows that the system could be operated under steady-state conditions with only a small amount of salt accumulation in the ducts and filter bags and with stack emissions of particulates, Resource Conservation and Recovery Act (RCRA) regulated heavy metals, and hydrogen chloride all within Environmental Protection Agency (EPA) allowable limits.

Flow measurements and a total mass inventory were taken for a 30-hour run, and a mass and energy balance was established for the 8-hour steady-state run. Overall, the test demonstrated 100.1 ± 5 percent mass balance based on the total solids analysis for the 8-hour test. The mass accountability for the total 30-hour operation, including system start-up, shutdown for an overnight test break, restart, and shutdown, was 98.7 ± 6 percent, based on total solids analysis. The individual mass balances for the evaporator and dryer during the 8-hour test period were 103.9 and 103.7 percent, respectively.

The mass balance for RCRA-regulated heavy metals showed 115 percent recovery of chromium and 192 percent recovery of lead. The apparently higher recovery of lead probably reflects difficulties in accurately measuring the low concentration of lead in the feed brine because there was no other source of lead in the system. The percentage of chromium recovered was within reasonable limits of uncertainty for trace metals. Chromium stack emissions averaged 69.5 percent of the EPA limit, and the maximum among three samples was 142 percent. Arsenic and cadmium concentrations in the brine and salts were below the uncertainty limit of measurement.

Average particulate matter stack emissions were less than 1.7 percent of the EPA limit of 30 mg/dscm

---

[1]Appendix C includes extract listings of the recommendations from the 1993 and 1994 reports prepared by the Committee on Review and Evaluation of the Army Chemical Stockpile Disposal Program (Stockpile Committee). The findings from the 1994 *Recommendations* report are also included. To assist the reader, an alpha-numeric code reference has been added to each finding and recommendation shown. These code references are applied to each use in the text.

(corrected to 7 percent O_2); stack emissions of hydrogen chloride were less than 1.8 percent of the EPA limit of 0.026 grams per second (0.206 lbs/hr).

During the past year, the Army also made a number of changes and conducted some acceptance tests at the TOCDF. Satisfactory completion of all the tests is required by the RCRA permit.

Changes in the TOCDF Brine Reduction Area Facility

The changes made for the TOCDF brine reduction area facility are described in the *Required Report for the Operational Verification Tests* (U.S. Army, 1993). The principal changes are outlined below.

Salt Particulate. JACADS experience showed that some salt particulate was escaping the dryers, causing overloading of the baghouses. The TOCDF has increased the processing rate by installing four pollution abatement system baghouses in the brine reduction area (versus two at JACADS) to handle the increased gas flow and particulate loading from the brine evaporators and dryers.

Condensate. At JACADS, there was condensate formation in the off gas ducting even though this gas was heated with an in-line burner to prevent condensation. Nevertheless, condensation occurred when the evaporator exhaust (which is almost 100 percent water vapor) was mixed with the dryer exhaust. A larger burner was installed at the TOCDF, and the evaporator exhaust was rerouted to enter downstream of the burner to keep gas from condensing.

Duct Work Leaks. Leaks, apparently caused by the accumulation of corrosive condensate, developed in JACADS duct work; the leaks were repaired by sealing and installing drain tubes. The TOCDF system was redesigned using a larger heater, a different configuration for introducing the evaporator exhaust, and sloping ducts. Although no condensation is expected, the design ensures that any condensate will drain back to the evaporator or to a knockout box.

Holes in Bag Filters. The bag filters at JACADS developed holes and had to be replaced. If the problem was caused by poor quality control of the bag fabrication or improper installation, the problem might recur at the TOCDF. For now, the TOCDF bags have been redesigned to be shorter (longer bags require a stronger pulse of air for cleaning than shorter bags).

Overfilling the Tanks. The high-level alarms in the brine storage tanks at JACADS malfunctioned, indicating that the tanks were full when they were not. The false alarm required the incinerators to be shut down to prevent the tanks from overflowing. The JACADS instrumentation was subsequently repaired, and improved instrumentation has been installed at the TOCDF. In addition, there are four 40,000-gallon storage tanks at the TOCDF (versus the two 26,000-gallon tanks at JACADS), which will allow greater flexibility before a tank-full signal would require incinerator shutdown.

Poor Drum Dryer Availability. The JACADS drum dryers proved to be unreliable, owing primarily to failures of the conveyor bearing and an inefficient conveyor wiper. The TOCDF design incorporates corrosion resistant conveyor bearings and a stainless steel wiper with a polyvinyl blade. It also includes inspection ports so that operators can check the wiper for salt buildup.

Acceptance Test Results

Certain portions of the brine reduction system were tested as part of the TOCDF systemization. Individual components (boilers, the pollution abatement system, surge tanks, evaporators, and drum dryers) were tested, and a systems material balance test was performed. Final stack emission tests have not been conducted because only simulated brine feed was available for testing. The results of testing should be similar to the successful JACADS runs (reported above) with modifications for the difference in processing rates.

The results of the Tooele material balance performance tests are reported in *Tooele Chemical Agent Disposal Facility, Phase 3 Systemization Demonstration Report, Brine Reduction Area Lines 1&2* (EG&G, 1994a). This report shows a discrepancy between the amount of salt collected and the theoretical amount of salt produced in the feed stream, i.e., the amount of dry salt collected represented only 80 percent of the predicted amount. However, a post-run inspection of the baghouses and condensate knockout boxes showed that few salts were present. This suggests that the discrepancy was primarily due to uncertainties in the salt content of the brine feed stream and the deposition of salt in the duct work. Operational procedures require monthly inspections and cleaning the duct work.

As a result of the above changes, the committee believes the test results are consistent with the successful demonstration of the brine reduction area at JACADS.

Dunnage Furnace

Satisfactory operation of the dunnage furnace and its related pollution abatement system was not demonstrated during the JACADS OVT. Therefore, the Stockpile Committee recommended that satisfactory operation be demonstrated prior to the start of agent operations at the TOCDF:

Demonstrate the dunnage furnace performance with various levels of chlorinated waste; if needed, either modify the pollution abatement system design (e.g., add acid gas scrubbing) or limit feed materials to those that can be handled by the existing design; alternatively, satisfactory land disposal options must be identified. (OVT2-2C)

The Army subsequently made a number of equipment modifications at JACADS and conducted the required RCRA trial burn tests during December 5–8, 1994. The following results of these tests were reported by the Army (U.S. Army, 1995b):

- The test feed stream consisted of cardboard boxes containing 260 pounds of wood dunnage along with polyethylene carboys containing 10.3 pounds of GB nerve agent absorbed in cellulose.
- No agent was detected (detection limit is 0.00006 mg/m^3) in the stack at any time during the four test runs.
- The particulate emission level averaged 6.7 mg/dscm, with a maximum concentration of 16.0 mg/dscm. This maximum concentration is less than 9 percent of the allowable rate of 180 mg/dscm and occurred during baghouse air flow pulsing (for bag cleaning) during which time the effectiveness of the "filter cake" on the bags was reduced from normal.
- The hydrogen chloride emission rate varied from 0.59 to 0.72 lb/hr, i.e., only 15–18 percent of the allowable 4 lb/hr limit.
- The unit was operated with a rolling average carbon monoxide concentration of 46.1 ppm in the exhaust gas downstream of the afterburner, well below the allowable 100 ppm level (one-hour rolling average on a dry basis corrected to 7 percent O_2), and with five-minute average concentration peaks of only 62.2 ppm, well below the 200-ppm five-minute peak limit.
- Separate composite leachate samples were made from furnace ash and baghouse ash. Toxic Characteristic Leaching Procedure (TCLP) analyses were performed for the eight TCLP metals as well as for additional metals. These analyses indicated that sufficient quantities of seven of the eight TCLP metals were present to warrant treatment of baghouse ash as a hazardous waste. Analysis of the furnace ash demonstrated sufficient quantities of four of the eight TCLP metals to require treatment as a hazardous waste.
- Dioxins and furans were both present in baghouse and furnace ash samples at the nanogram per kilogram level. There are no regulatory limits on dioxins/furans concentrations in ash.
- Bis (2-ethylhexyl) phthalate was the only semivolatile compound found in the ash samples; it was detected (1,640 ng/kg) in one sample of furnace ash. No semivolatile compounds were found in the baghouse ash. No volatile compounds were detected in either ash sample.[2]
- The results of the following measurements were in compliance with EPA standards:

    a. Exhaust-stack emissions including: O_2 (oxygen); CO_2 (carbon dioxide); volatile and semivolatile products of incomplete combustion; polychlorinated dibenzo-p-dioxins and dibenzofurans (PCDD/F); trace metals; particulate breakthrough; total hydrocarbons; NO_x (nitrogen oxides); SO_x (sulfur oxides); HF (hydrogen fluoride);
    b. Furnace scrubber liquid process samples for: agent GB; volatile and semivolatile products of incomplete combustion; polychlorinated dibenzo-p-dioxins and dibenzofurans; pH (acidity or alkalinity); trace metals; and reactivity.

---

[2]Waste materials will be tested routinely. Toxic liquids can be disposed of at a hazardous waste facility through deep well injection and toxic solids in landfills.

Modifications to the Dunnage Furnace

The principal changes to the TOCDF dunnage furnace pollution abatement system (U.S. Army, 1993), are described below.

Temperature Control. The temperature of the JACADS dunnage furnace was to be controlled by limiting inlet air and causing oxygen-deficient combustion. However, air leaking in through the furnace-door seals allowed the dunnage load to burn at an uncontrolled rate. The JACADS unit was modified to operate with an excess of air by controlling the fuel content of the feed batch rather than the air intake. The TOCDF unit was redesigned to operate in a similar way.

Induced Draft Fan Failure. Operating difficulties were experienced with the JACADS dunnage furnace after the induced draft fan failed with dunnage fuel still in the primary combustion chamber; the failure of the fan also locked out the furnace burner fuel supply. At both JACADS and the TODCF, backup emergency induced draft fans have been added, as well as cooling water spray nozzles located in the top of the primary combustion chamber. The spray cools the dunnage fuel, generating steam that helps purge the chamber. The induced draft fan and the water spray system are designed to keep negative pressure in the system and to keep the combustible gas concentration below the lower explosive limit during cooling of the dunnage fuel.

Lift Shaft Fire. On one occasion the JACADS dunnage furnace ram was unable to push into the furnace a box of dunnage that had caught on the edge of the furnace door. Radiant heat from the furnace ignited the box in the entry shaft. The TOCDF dunnage furnace feed system has been redesigned with a higher ram pushing force and a shorter cycle time. In addition, a new heat shield reduces preheating of the box before the furnace door is completely open.

Lift Shaft Hydraulic Fluid. Furnace heat caused the ethylene glycol/water hydraulic fluid used at JACADS to thicken gradually. This eventually caused the lift to bind. Both JACADS and the TOCDF will now use FYRQUEL 220, a high-temperature hydraulic oil. Also, the lift will be cycled several times a day during furnace operations to keep the lift seals and shaft lubricated properly.

Ram Failure. The hydraulic ram on the JACADS dunnage furnace failed to retract on one occasion, damaging bearings and instrumentation. The TOCDF dunnage furnace ram system has been completely redesigned. The original JACADS hydraulic unit has been replaced with an electrical powered unit, and a heat shield has been added to protect the ram during door cycling.

Fuel Oil Strainers. JACADS fuel oil strainers frequently became plugged up, requiring unscheduled maintenance. The furnaces at the TOCDF will burn natural gas rather than fuel oil, thus eliminating this problem.

Baghouse Pressure Drop. The JACADS RCRA permit requires that the baghouses operate with a pressure drop of 1–20 inches of water column. Higher pressure might cause bag plugging, and lower pressure might indicate that the bags are torn or leaking. There were problems at JACADS maintaining a minimum pressure drop because the combustion rate and exhaust gas flow rate were lower than expected. However, periodic visual inspection revealed no problems with the bags, and the permit was modified to allow for a lower pressure drop at reduced operating rates. The TOCDF flow rates and bag sizes will differ from those at JACADS. However, operational control over the dunnage feed rates will continue to be required in order to maintain a proper pressure drop.

Pressure Excursion. A maintenance operator inadvertently shut off the emergency quench water spray at the JACADS dunnage furnace during an incineration operation, also locking out the furnace fuel feed burners. Continued burning of the dunnage load in the furnace (operating in an oxygen-deficient atmosphere) then produced combustible gases that migrated into the lift closure where they ignited, causing overpressurization that damaged the shaft and surrounding wall panels.

Both JACADS and the TOCDF have since added a water (steam) purge to the primary combustion chamber to cool dunnage fuel and to keep the combustible gas concentration below the explosive limit in case the burners lock out while the dunnage load is still burning.

Dunnage Furnace Nonagent Test Results

Two major nonagent tests were conducted at the TOCDF (EG&G, 1994b, 1994c). These tests were not intended to be performance acceptance tests, which will be conducted after agent processing begins.

A thermal capacity test of the modified TOCDF furnace was conducted to demonstrate the following capabilities: (1) the capacity to handle dunnage disposal at the required rate; (2) operability of the furnace and related pollution abatement system in automatic mode; and (3) the operability of various components, including the dunnage furnace pollution abatement system. Over a continuous eight-hour period the tests demonstrated a thermal capacity to dispose of 431 lb/hr of mixed dunnage as compared to the original design capacity of either 1,000 lb/hr of wood dunnage or up to 24 mine drums (560 pounds of metal) per hour.

The Army maintains that the dunnage furnace is an optional part of the overall disposal operation and is primarily used to minimize waste disposal costs. There is also a program to minimize the feed stream to the dunnage furnace. Nevertheless, improving the operating efficiency of the TOCDF dunnage furnace is important because it is currently designated as the principal system for reducing facility hazardous waste.

A water spray purge test was performed to determine if the emergency fan, combined with water spray purge cooling in the event of a primary induced draft fan power failure could maintain a negative pressure in the primary reaction chamber. Each of the three test runs was successful, indicating that water spray cool-down could be used, along with the emergency induced draft fan in the event of a failure of the main induced draft fan and a lockout of furnace burner fuel supply. During tests with 200-pound and 500-pound dunnage loads, furnace pressure remained negative, no combustible gas concentrations were detected, and carbon monoxide concentrations remained below RCRA limits (<100 ppm) at all times (EG&G, 1994c). In the final analysis, the level of exhaust gas emissions from the modified dunnage furnace at the TOCDF is likely to be similar to or lower than the level of gas emissions at JACADS.

Overall Assessment of Dunnage Furnace

After reviewing the changes and tests described above, the committee believes that the dunnage furnace will perform acceptably at the TOCDF.

Nitrogen Oxide Emissions

During JACADS OVT operations, nitrogen oxide (NO_x) emissions during VX destruction were acceptable, but the Stockpile Committee recommended that NO_x emissions be analyzed with respect to more stringent requirements that might apply to U.S. sites:

Review the probable levels of NO_x production from VX destruction and the allowable emission levels at the other continental U.S. sites requiring VX destruction; if appropriate, develop needed NO_x abatement systems. (OVT2-2D)

There are currently two general regulatory standards to be met for nitrogen oxides: (1) the prevention of significant deterioration of the air quality which sets a limit of 250 tons per year of NO_x (based on a 12-month rolling average) at any single facility; and (2) the requirement that the NO_x concentration not exceed 1 $\mu g/m^3$ at the facility boundary, again based on a 12-month average. If calculations show that either of these limits may be exceeded, then the facility must demonstrate by modeling that the excess NO_x will not damage the local air quality. If the modeling demonstrates a deterioration of air quality, NO_x controls must be added to the air pollution abatement system.

The Ralph M. Parsons Company (Parsons, 1994) has estimated NO_x emission levels for the TOCDF using an equilibrium model to calculate mass and energy balances on the various furnace systems comprising the baseline system. Because NO_x formation is kinetically controlled, the equilibrium model may predict much higher thermal NO_x emissions than will actually be produced. (Note: This was confirmed during testing of the TOCDF.) The thermal capacity tests at the TOCDF during systemization provided NO_x test data that can be used to develop an empirical relationship (equation) between NO_x emissions and furnace temperature for the liquid incinerator, the deactivation furnace system, and the metal parts furnace. This equation makes possible the prediction of NO_x emissions, as a function of temperature only, for furnaces from any system. The resulting equation was used to adjust NO_x concentrations predicted by the Parsons equilibrium model and energy balances. Because no munitions are being burned in the tests, and the fuel is natural gas, the only nitrogen source is the combustion air. When nitrogen-containing constituents are combusted in baseline system furnaces, the NO_x levels will be higher. Currently, there are no data for "fuel NO_x" because the TOCDF has not started processing munitions.

The worst case scenarios for annual NO_x emissions were calculated for each of the eight continental U.S.

sites (see appendix B for stockpile locations). First, the total contribution of NO_x emissions was calculated for each incinerator, burner, and boiler using only fossil fuel. Second, the total contribution of agent (fuel) NO_x emissions was calculated using design processing rates for each incinerator and a worst case processing schedule. (Note: Calculations are based on operations of 6,000 hr/yr and furnace idle time of 2,760 hr/yr.) The principal source of agent fuel NO_x is in the VX feedstock. The Stockpile Committee believes the incineration of VX should not significantly increase the Parsons estimates for the following reasons:

- There are relatively small quantities of VX in the stockpile.
- Annual NOx emissions are regulated by 12-month rolling averages; short VX campaigns will be averaged with non-VX campaigns during any 12-month period, thus reducing the impact of processing VX.
- The estimates for generating NO_x from VX incineration, although not as conservative as the estimates for non-nitrogen-bearing munitions, still assume a significant conversion efficiency of agent fuel nitrogen to NO_x.

The Parsons mass and energy balances indicate that none of the sites will exceed the limit of 250 tons of NO_x emissions per year. Calculation of the point dispersion impact (perimeter concentration levels) are site-specific and contingent on local meteorological conditions as well as the distance to the facility boundary. Point dispersion calculations have not been completed for all sites owing to the unavailability of site-specific meteorological data, but the Stockpile Committee believes that NO_x emissions will not cause a major impact at any location because of the relatively small amount of NO_x expected to be generated.

The main stack at each facility will include an NO_x meter and a total exhaust gas flow meter. The actual NO_x emissions both during furnace idle (not processing) and during agent operations (including trial burns) will be recorded and tracked against the estimates.

Liquid Incinerator Slag Removal

The combustion of agent in the liquid incinerator converts most of the carbon and hydrogen to volatile gases. However, inorganic elements are changed to acids, oxides, or salts, which condense on the walls of the slightly cooler secondary combustion chamber to form molten slag that slowly flows down the walls and collects in a pool at the bottom of the chamber. (The TOCDF slag production rate will be on the order of 100 pounds per hour, depending on the agent being destroyed.)

Because the accumulated slag must be removed periodically, the JACADS liquid incinerator was designed with this in mind, using principles that were then being developed and tested at the Chemical Agent Munitions Disposal System (CAMDS) facility. These involved the use of a sloped furnace chamber bottom with a central tap hole from which the liquid slag could be drained. However, subsequent CAMDS tests (after construction of JACADS was begun but prior to JACADS start-up) showed that the system would not operate effectively because the slag solidified. Consequently, the liquid drain system was never used or tested at JACADS.

Periodic shutdown of the system to allow operators to enter the secondary chamber and remove the hardened slag manually has contributed significantly to downtime of the JACADS liquid incinerator. Based on the JACADS experience, shutdown and manual removal procedures would limit availability of the TOCDF liquid incinerator to only 30 percent of peak design throughput. The Stockpile Committee recognized this as a significant problem:

> Develop and demonstrate the proposed hot-slag removal system for the liquid incinerator system. (OVT2-2E)

To minimize this problem, a program was initiated to develop a new, heated, liquid slag removal system for installation at the TOCDF and other plants in the continental U.S. The complete system is referred to as the slag management system. This development program initially involved a review of existing commercial methods and equipment for removing slag in related industrial applications. A specific design was then developed from these various alternatives. Because this design was not completed in time for demonstration at JACADS or implementation at the TOCDF prior to the start of systemization tests, the committee recommended that the demonstration be completed prior to the start of agent operations at the TOCDF.

Once it became clear that the JACADS design would not work, the TOCDF furnaces were redesigned using a flat bottom instead of a sloped bottom. The bottom was elevated so that subsequently designed equipment could be installed under it.

The new slag management system design involves cutting a central hole in the existing flat furnace bottom, attaching a cylindrical extension of the furnace under the central hole, and providing a side tap system from this extension to drain molten slag at intervals during operations.

The side tap system involves a vertical slide valve to stop slag flow, a remotely operated external drill to drill through the frozen slag into the inner molten zone, and electrical resistance heaters to keep the drilled channel molten during periodic drainage operations. The heaters are turned off after the drainage operations are completed, allowing the outer layer of slag to solidify because the solid is less corrosive.

At peak operating rates, slag drainage will be necessary about once a week. Slag draining operations will be conducted with the furnace hot and operating, but agent feed will be stopped to avoid any possibility of agent escape with the slag.

The newly installed slag management system at the TOCDF will be mechanically tested prior to the start of agent destruction operations. However, because of the variability of the slag, the Army decided that there was no adequate surrogate for complete testing of the system, which will be tested only when slag accumulates from agent destruction operations. If the slag removal system does not allow for hot slag removal, the current manual removal procedure will be resumed. The committee will complete assessment of the slag removal system after testing with agent at the TOCDF.

Furnace Feed System

The Stockpile Committee recommended that better methods for tracking various types of munitions be implemented to avoid furnace feed errors of the kind encountered during OVT testing at JACADS:

> Eliminate furnace feed errors by improved monitoring and control of the deactivation furnace and metal parts furnace feed systems and by improved methods for tracking the various types of munitions. (OVT2-2F)

The Army responded to this recommendation by retaining the MITRE Corporation to assess munition tracking problems identified at JACADS and to provide appropriate recommendations for resolving munitions tracking problems that have been identified.

The MITRE Corporation report (MITRE, 1994) presents a number of recommendations for modifications at the TOCDF. In addition to the problems observed from JACADS, MITRE identified 13 other process limitations that might compromise the effectiveness of tracking munitions. MITRE concluded that the enhancements implemented for the TOCDF rocket processing line and the projectile processing line should be adequate to minimize the likelihood of repeating munitions tracking failures during future chemical agent disposal facility operations.

MITRE's analysis concentrated on each processing campaign (rockets, projectiles, bulk items) separately and on the critical steps in each campaign. JACADS OVT malfunctions that interfered with the successful completion of the critical step were analyzed, and recommendations for correcting the deficiency were proposed. Additional improvements were also recommended to the Army for consideration.

The following changes for the rocket processing line have been implemented at the TOCDF:

- To ensure the full drainage of rockets, an interlock has been added that requires a signal from sensors in the agent quantification tank to confirm that rockets have been properly drained before the rockets can be moved out of the drain station.
- An interlock constraint on feed rate of rockets has been added to the deactivation furnace to prevent loading in excess of that allowed in the RCRA permit.

Changes for the projectile processing line include:

- Fail-safe proximity sensors ensure that the trays of projectiles remain on the tray conveyor system and that trays cannot proceed without this verification.
- The pick-and-place loader primary control has been moved from local control to the central control system. The local control instrument did not have the capability of monitoring manual operations. The control room monitors both instruments and manual operations.
- An interlock constraint on the feed rate of projectiles to the metal parts furnace has been added to

prevent loading in excess of that allowed in the RCRA permit.

Changes for the bulk item processing line include:

- A bubbler device has been added to indicate the level of agent remaining in a munition or bulk container.
- An encoder on the bulk drain tube may be installed to determine the level of liquid left in the container. With improvements in operations of the load cells, this might be enough to ensure the proper drainage of containers. The Army is considering this improvement, and tests are being run. In addition, the Army is developing permit terms with the Utah Division of Solid and Hazardous Waste (DSHW) to deal with containers that cannot be fully drained because of gelled agent.
- An interlock constraint on the feed rate of bulk items has been added to the metal parts furnace to prevent loading in excess of that allowed in the RCRA permit.

Changes to prevent problems from the manual override of interlocks include:

- Adding sensor fault trapping, which indicates to the control system when changes in critical sensors or interlocks occur.

The implemented improvements have been satisfactorily tested during integrated plant runs as part of the TOCDF systemization process (MITRE, 1995). The changes made and tested at the TODCF address the problems identified during the JACADS OVT and should provide for substantially safer furnace feed performance than at JACADS.

Residual Gelled Agent

Determining the residual agent level and detecting residual gelled agent in ton containers and spray tanks (bulk containers) is essential to establishing a remote real-time indication of agent level while a container is being drained. The method of measuring the level of residual agent during JACADS operational verification testing required that personnel wearing demilitarization protective ensembles (DPEs) manually insert a dipstick into each container. An accurate measure of residual agent in bulk containers is necessary to ensure compliance with RCRA permit requirements. Also, the residual level is critical because it affects the processing rate of the metal parts furnace. Consequently, the Stockpile Committee recommended that this problem be addressed:

> Address all problems associated with residual gelled mustard, in particular, the use of suited personnel to perform functions that were intended to be automated. (OVT2-2G)

Since completion of JACADS OVT, the agent collection and quantification system at the bulk drain station (BDS) for the TOCDF has been modified (GPS Technology, Inc., 1993). This modification was tested at the Chemical Demilitarization Training Facility, and the finalized design was implemented at the TOCDF. The TOCDF team is also updating operating procedures to incorporate revisions in the agent collection and quantification system into the process operating procedures.

Modifications to the bulk drain station emulate the accepted design for the residual agent level sensing system of the multipurpose demilitarization machine. The modifications to the bulk drain station fulfill several other remote processing needs at JACADS. Measuring the residual agent level inside a bulk container is facilitated by a bubbler orifice at the lower end of a new suction tube inserted into the container. This presents a remote real-time indication of agent level while the suction tube is still in the bulk container at the bulk drain station, thus providing the operator with a discrete go/no-go reading. The sensor and connected instrumentation are remotely and automatically checked at the beginning and the end of each operation to ensure that the system is operating properly. This eliminates the need for manual dipstick measurements of agent remaining in a bulk container after drainage. If the bulk drain station is not in use, or an area ventilation system upset occurs, the system is automatically disabled and isolated.

The systems for sensing the residual level of agents in bulk containers also alert operators to "pump out" problems, which are indicative of residual gelled agent. These systems should help to increase plant processing performance to the designed capacity, and the remote-automatic measurement procedure should substantially reduce the number of entries by personnel wearing demilitarization protective ensembles.

Environmental Permitting and Regulatory Requirements

With respect to environmental permitting and regulatory requirements, the Stockpile Committee recommends that the Army:

> Establish and maintain close working relationships with permitting agencies, and support these efforts with careful analysis of operating parameters to ensure that permits provide for safe destruction of agent, adherence to regulatory requirements, and effective plant operations. (OVT2-3)

The TOCDF staff has established a set of comprehensive activities to ensure timely and effective coordination with the Utah DSHW. This coordination has yielded positive results in identifying problems related to environmental permitting and compliance and in developing solutions to these problems. Some examples of these coordination efforts are:

- *Technical responses.* A quick response system has been established to respond to DSHW questions and requests for information. Typically, the response is provided by a member of the TOCDF team making a personal appearance at the DSHW offices in Salt Lake City. For example, during the review of the liquid incinerator #1 surrogate trial burn plan, daily meetings were held each afternoon to respond to comments and questions resulting from the DSHW review. Because of this level of coordination, the liquid incinerator #1 surrogate trial burn plan was completed on schedule. This approach is being used for other document reviews as necessary.
- *Planning meetings.* Planning and scheduling meetings attended by the TOCDF team and the DSHW were held on a monthly basis until March 1995. As the schedule became tighter, meetings were held approximately once a week. These planning and scheduling meetings are used to prioritize the efforts of both the TOCDF team and the DSHW personnel.
- *TOCDF presence at the DSHW offices.* In addition to face-to-face coordination meetings, members of the TOCDF team are present at the DSHW offices at least two days a week to help develop DSHW work schedules coordinated with TOCDF schedules, and to ensure that the documentation in the DSHW information library is current.
- *DSHW presence at the TOCDF site.* The DSHW established an office at the TOCDF site, which is staffed at least two days each week, and at other times as necessary (e.g., for the completion of required inspections). Once agent operations start, it is anticipated that the DSHW will maintain a presence on site.
- *Training of DSHW personnel in chemical demilitarization.* Representatives from the DSHW attended a training course (a short version of the control room operator course) at the Chemical Demilitarization Training Facility. After the course, DSHW personnel had a better understanding of how the systems work and, more importantly, why they work the way they do. The comprehensive sets of activities outlined above should ensure the safe destruction of agent, adherence to regulatory requirements, and effective plant operations.

Environmental Compliance

The Stockpile Committee believes that environmental compliance and the Chemical Stockpile Disposal Program should go hand in hand. This belief is reflected in the committee's clear recommendation regarding compliance:

> Establish programs, procedures, and management oversight to ensure continuing compliance with all environmental regulations. (OVT2-4)

EG&G Defense Materials, Inc., the Army's implementing contractor, has the overall responsibility for ensuring that the TOCDF is in compliance with all applicable federal, state, and local environmental requirements. To meet this responsibility, EG&G has drafted an Environmental Compliance Plan (ENVCP) (EG&G, 1995a), a summary document that provides guidelines for general environmental compliance, background for the environmental impact statement, and an overview of the environmental requirements to which the TOCDF will adhere. The ENVCP is a flexible document that is modified, subject to regulatory approval, to accommodate specific phases of the project, e.g., systemization, operations and decommissioning, and changing environmental regulations affecting specific phases of the project.

To ensure timely and complete implementation of the ENVCP, the following roles and responsibilities have been assigned:

- The Environmental, Safety, and Surety Manager is responsible for assuring implementation of all environmental procedures by the Environmental Compliance Group, including the maintenance and updating of all environmental procedures; notifying regulatory agencies and management personnel of EG&G Defense Materials, Inc. of major reportable deficiencies and completing all paperwork required by company, state, and federal rules; informing EG&G management of compliance status and trends.
- The Environmental Compliance Coordinator ensures that both manpower and material assets are available to execute all environmental inspections; ensures that reports are drafted on time and that follow-up inspections are executed for *all* deficiencies; is responsible for administering the TOCDF Environmental Assessment Program, a periodic evaluation—via audits, surveillance, or inspections—of TOCDF compliance with federal, state, Department of Defense, and Tooele Army Depot environmental permits, rules, and regulations.
- Environmental Inspectors oversee and provide technical assistance to operators of the regulated areas; conduct environmental compliance inspections in accordance with written procedures.
- Quality Assurance/Quality Control Manager assists the environmental manager with overall guidance for the proper implementation of the Internal Environmental Audit Program, including auditing hazardous waste transporters; ensures that the appropriate personnel are properly trained, qualified, and certified in auditing techniques and procedures and that environmental audits of hazardous waste transporters are planned, executed, and documented.
- All TOCDF employees report noncompliance with environmental requirements to their supervisors, no matter how insignificant an incident may appear. Reports are forwarded to the environmental office for evaluation against reportable event criteria. Representatives from the office of the Program Manager for Chemical Demilitarization (PMCD) are then notified; required notifications to the Tooele Army Depot or the Utah Division of Solid and Hazardous Waste are coordinated by the PMCD; telephone notification is then made, followed by a written report.

Operational procedures for environmental inspections have been established for the following areas: the munitions demilitarization building; the personnel and maintenance building; the container handling building (CHB); the monitor support building; the chemical assessment laboratory; the outbuildings at the TOCDF; the residue handling area; the brine reduction area; the pollution abatement systems; TOCDF satellite accumulation sites; the process and utility building; and the 90-day hazardous waste storage sites. In addition, environmental procedures have been developed for hazardous waste transport audits; environmental self-assessment; environmental inspections; and incorporating environmental compliance requirements into operating procedures, project regulatory procedures, and drawings.

The ENVCP ensures compliance with all environmental laws, rules, permits, and regulations by identifying shortfalls in environmental/operational design and execution. The environmental compliance activity accomplishes this through daily surveillance inspections and periodic audits of both on-site regulated activities and off-site treatment, storage, and disposal facilities that receive waste generated at the TOCDF. Databases to track nonconforming conditions are maintained and periodically examined to identify significant problems. Procedures are reviewed, and changes or new regulatory requirements are incorporated and "tagged." Opportunities for hazardous waste minimization and pollution prevention are also investigated. Implementation of the ENVCP and strict adherence to the established environmental procedures listed above should ensure continuing compliance with all environmental regulations.

Overall Safety

One major design improvement at the TOCDF over JACADS was the better incorporation of human factors in the design of the physical plant. Many of the poorer design aspects of JACADS, such as inaccessible manual valves and meters, especially in areas requiring workers wearing demilitarization protective ensembles, have been eliminated at the TOCDF. These design problems had been implicated in upsets at JACADS, such as operators not monitoring an agent flow gauge that was difficult to reach. The Stockpile Committee reacted in part to this situation by making the following recommendation:

Develop systems to improve overall management of safety. (OVT2-5)

At the TOCDF, the committee noted that most places requiring manual access were within reasonable human reach and sight. The pipe work was clearly labeled as to flow direction and contents; good use was made of color coding of temporary connections; and computer console displays and video surveillance systems in the control room were designed to be more user-friendly. The workers interviewed at the TOCDF commented favorably on the quality of these improvements.

Changes Resulting from Risk Assessment

At the following recommendation of the Stockpile Committee, the Army undertook a site-specific quantitative risk assessment (QRA) at the TOCDF:

Complete the risk assessment for the Tooele Chemical Agent Disposal Facility during the systemization period. (OVT2-6)

Although the assessment for the TOCDF has not yet been completed, routine reviews of information from the risk assessment for campaigns 1 and 2 (SAIC, 1995a) have led to changes in the facility and in its schedules to reduce risk. These are described further in chapter 6, "Overview of Site-Specific Risk Assessment." Specifically, the following four changes are currently being implemented:

- Operation of the metal parts furnace feed airlock. The QRA suggested that flammable vapor might accumulate in the metal parts furnace feed airlock, especially for bulk agent containers. Venting the airlock to the furnace afterburner is being evaluated, as are procedural changes to limit the residence time for items in the airlock. A solution will be found prior to the start of agent operations in the metal parts furnace.
- Weteye bomb aluminum and agent interaction. The QRA identified the potential for interaction between molten aluminum and liquid agent in the metal parts furnace during processing of aluminum weteye bombs. Interaction could cause an explosion within the furnace. As a result of this finding, the order of the campaigns was changed. Ton containers instead of weteye bombs will be co-processed with GB rockets during the first campaign. Further review of the Science Applications International Corporation (SAIC) calculations and development of processing strategies that will avoid molten aluminum and agent interaction (SAIC, 1995a) are under way.
- Weteye bomb handling and inventories. The QRA found that weteye bombs are significant contributors to risk because they contain GB and are relatively thin-walled. Additional analysis in the QRA may indicate the number of weteye bombs that should be stored in the container handling building to minimize risk.
- Seismic anchorage of the liquid propane gas tank. The facility review of equipment fragilities during an earthquake indicates that the liquid propane gas tank anchorage has some seismic vulnerability despite its construction to seismic zone three requirements. The information provided by the QRA team will be used as a basis for evaluating the need for additional bracing for this tank.

Risk reduction from these changes is a direct result of the risk assessment. The committee anticipates that additional improvements will be identified and implemented as the TOCDF risk assessment is completed early in 1996.

REVIEW OF NRC RECOMMENDATIONS REGARDING THE MONITORING SYSTEM AT JACADS

After reviewing the monitoring systems for the destruction of chemical agents and agent by-products at the JACADS during systemization, the committee issued a report entitled *Review of Monitoring Activities Within the Army Chemical Stockpile Disposal Program* (NRC, 1994b). This report included five general recommendations and ten specific recommendations. The specific recommendations addressed six issues involving plant-wide agent monitoring, and exhaust stack agent and agent destruction by-product monitoring, and four issues affecting the operation of the analytical laboratories supporting both agent and nonagent monitoring activities. These recommendations were motivated by the finding that "the monitoring system currently in use at JACADS should be improved prior to employment at sites in the continental United States" (NRC, 1994b). In this section the recommendations presented in the *Monitoring* report are repeated and the Army's responses to these recommendations are reviewed and evaluated.

In response to the general recommendations, the Army has addressed monitoring issues for both agent and agent destruction by-products more attentively and actively than was evident before the *Monitoring* report (NRC, 1994b) was issued. In general, when specific issues could be addressed with improved deployment of, or incremental improvement to, standard agent monitoring techniques or utilization of other commercially available analytical instrumentation, the Army has done a very good job. However, when the implementation of committee recommendations required significant research and development to create or identify new technology, the Army's response has been tentative and less effective. This is partially attributable to the fact that the Chemical Stockpile Disposal Program is a major defense acquisition program attempting to slow cost growth attributable to previously unbudgeted requirements. In addition, the agent monitoring program did not have experience in procuring and managing research and development and has been slow to master these new skills.

General Recommendations for Agent/Nonagent Monitoring

In the *Monitoring* report, the Stockpile Committee made five general recommendations. The first was:

> The Army should initiate a substantial program to upgrade the monitoring systems for continental U.S. sites. (MON-1)

In response, the Army tasked senior personnel within the Environmental and Monitoring Division of their Chemical Demilitarization and Remediation Activity (now Program Manager for Chemical Demilitarization) to initiate and sustain a substantial program to upgrade monitoring systems for the U.S. sites, commencing with the TOCDF.

The second general recommendation was:

> The Army should obtain expert help at both the systems design and the equipment selection levels, perhaps by engaging a contractor with extensive experience in monitoring of trace species and in advanced instrument development. (MON-2)

The Army has accordingly retained contractors to help evaluate and select upgraded monitoring equipment (SAIC, 1995b); however, the contractors involved are generally competent only to assist in identifying standard, commercially available analytical and monitoring equipment. They lack the background necessary for deciding whether some new monitoring technology, which is only available for research, should be improved or adapted to meet the needs of the disposal program.

The third general recommendation was:

> The Army should undertake whatever instrument development is necessary to ensure that improved instrumentation is available to the chemical disposal program in suitably rugged and operational forms. (MON-3)

In response to specific committee recommendations, the Army has let one instrument development/demonstration contract for a basic agent monitoring technology (Fourier Transform Infrared system) that is not currently available commercially. However, additional research and development, either by the Army's laboratory programs or by outside contractors, has not been initiated yet so that additional technologies that might provide longer term upgrades in monitoring capabilities can be tested and promoted.

The fourth general recommendation was:

> The Army should test and use new monitoring instrumentation at JACADS before such instrumentation is employed at Tooele. (MON-4)

The Army has initiated both new deployment modes of existing monitoring technology and deployment of new commercial analytical instrumentation at JACADS to evaluate their effectiveness for use at the TOCDF.

The fifth general recommendation was:

> The Army should plan to continually improve the monitoring system in areas where performance is presently limited by unavailability of suitable instrumentation. (MON-5)

The Army has initiated activities to redress most of the deficiencies that have been identified in monitoring capability. The Stockpile Committee believes that the Army is committed to the continuous improvement of monitoring systems at the TOCDF and other U.S. disposal sites.

Specific Recommendations for Agent/Nonagent Monitoring

In the *Monitoring* report, the Stockpile Committee made ten specific recommendations, of which the first six addressed agent/nonagent monitoring. The first recommendation in this group was:

> Add the capability for positive identification of chemical agent species (chemical speciation) to the agent detection system and analytical laboratories at all of the disposal facilities in order to reduce the occurrence of false positives. (MON-6)

The present analytical system uses gas chromatography, which is sensitive to many chemically similar compounds leading to false positive signals. A single false positive requires shutdown of agent operations but does not by itself initiate the response appropriate for a major agent release. False positive signals disrupt plant operations and increase the potential for human error and equipment degradation. Using mass spectrometric detection in the analytical laboratory now provides improved discrimination between chemical species, as would using the next generation of reliable, rugged infrared spectroscopy-based field instruments with adequate detection limits throughout the plant.

In response to the above recommendation, the Army has deployed gas chromatographic systems with mass spectrometric detectors (GC/MSD) at selected sites at JACADS and the TOCDF to identify the specific chemical species responsible for activating the Automatic Continuous Air Monitoring System (ACAMS) agent alarms. The new instruments are deployed in the analytical laboratories at each site and supplement the gas chromatographic/flame photometric detector (GC/FPD) analysis of Depot Area Air Monitoring System (DAAMS) absorption tube samples. These samples are used to determine whether an ACAMS alarm was triggered by agent release or was a false positive alarm triggered by another chemical. Procedures have been devised and tested for GC/MSD to analyze an additional DAAMS sample with every ACAMS alarm. This procedure facilitates the identification of the specific chemicals that frequently cause false positive ACAMS alarms. Once they have been identified, these chemical can be eliminated from the disposal facilities, thus reducing the level of false positive alarms.

In addition, the Army is planning a field test in the analytical laboratory at the TOCDF of a gas chromatographic system with both mass spectrometric detectors and atomic emission detectors (AED). This GC/MSD/AED system ought to provide more information about the identity of substances that trigger false positive alarms. The Army plans to test this instrument at the TOCDF in June 1996 with various potential trigger substances. A more species-specific alarm to complement the current ACAMS system needs to be developed for both general plant and stack monitoring.

The second specific recommendation was:

> Institute continuous monitoring for all agents present at each facility, including those in storage areas. (MON-7)

With single-agent monitoring systems, there is a risk that release of a different agent from mislabeled munitions or leaky storage containers might go undetected. Single agent monitoring systems also increase plant downtime during agent changeover operations. There are several ways to conduct multiple-agent monitoring, including the simple installation of additional ACAMS at each site. These modifications should be weighed against possible increases in the false alarm rate.

The Army's deployment of ACAMS set at the time-weighted average (TWA) exposure level for each agent (GB, VX, HD) in the agent unpack area at each site is a positive step. ACAMS for each agent are also deployed on the ventilation air carbon filter units and on the common incinerator exhaust stack. The Army has announced plans to implement a similar deployment in the agent storage areas as recommended by the committee.

The third specific recommendation was:

> Reduce the time required for confirmation of false positives. (MON-8)

Confirming ACAMS alarms requires retrieval and laboratory analysis of DAAMS sample tubes. The manual transport of tubes and the analysis can lead to delays of 15 to 20 minutes. Analysis times can be reduced by using analytical methods with greater agent specificity. But, manual transport delays, a significant component of overall delays, will still be a problem.

A significant reduction of the time required to confirm ACAMS alarms would require development and deployment of a rapid, species-specific alarm system alongside each ACAMS. At present, the Army has not yet identified monitoring technology to fulfill this function. However, the Army has initiated a development program for an advanced electronics architecture, dual detector technology ACAMS, which should provide agent detection with a three to five minute response time as well as detection redundancy.

The fourth specific recommendation was:

> Evaluate the procedures for periodic testing of field sensors to ensure that false negatives are not possible if a significant release should occur. (MON-9)

Potential failure modes in sample collection could produce false negative responses. Monitoring system challenge tests are now conducted, but a statistical analysis of test results could be used to identify common failure modes. Once these are identified, changes in monitoring equipment or its maintenance could minimize the number of false negatives.

In briefings to the committee, the Army has indicated high confidence that its procedures for regular field challenges of ACAMS and DAAMS field monitors using agent samples in calibrated solutions are adequate to eliminate a significant number of false negatives. The current field challenge procedures were developed in 1991 in collaboration with the Department of Health and Human Services (DHHS). The Utah Department of Environmental Quality (DEQ) has questioned the extent to which ACAMS agent challenges test the air sampling component and the GC/FPD detector portion of the ACAMS system.

The current field monitor challenge procedures test several levels of calibrated agent solutions on a daily basis. These procedures appear to be effective in identifying failing ACAMS. The Army must continue to analyze the results of these tests to identify common failure modes and to institute maintenance procedures to minimize field failures.

The fifth specific recommendation was:

> Implement monitoring designed to provide more rapid response to high-level agent release. (MON-10)

The Army is planning to deploy time-phased (staggered) ACAMS on the common incinerator exhaust stack at both JACADS and the TOCDF. This is a good first step toward an alarm system that will cut the alarm time in half in response to a major release.

The response time of the standard ACAMS instrumentation lengthens when the ACAMS is operated at greater sensitivity. An ACAMS set for low detection levels with longer sampling times might lead to unnecessary delays in response to high levels of agent release. Supplementing low-detection-level ACAMS with high-detection-level ACAMS may be an acceptable way of improving detector response times.

This recommendation and the Army's response are closely coupled to recommendations MON-8 and MON-9. To date, the Army has committed to deploy ACAMS in the unpack area at both JACADS and the TOCDF set at the high immediately dangerous to life and health (IDLH) level. These ACAMS will respond more rapidly than ACAMS set at normal levels. In addition, time-phased (staggered) ACAMS are being deployed in the common stack for faster "stop feed" control of furnaces. These systems will also be in place at the TOCDF before the start of agent operations. Both the Stockpile Committee and the Army recognize that new monitoring technology will be required to reduce the desired response times to a few seconds, rather than the few minutes possible with staggered ACAMS. The Army has contracted for the development and demonstration of a Fourier Transform Infrared (FTIR) multipass absorption technique, which should be capable of real-time (~1 second) detection of high agent release levels.

The sixth specific recommendation was:

> Evaluate the benefits of more frequent analysis of facility stack gases for nonagent trace contaminants. (MON-11)

Daily analysis of samples of flue gas for products of incomplete combustion, particulates, and metals may provide an important database to monitor incinerator operating conditions and the environmental impacts of the disposal system. Although analysis of these emissions is required only in trial burns, more frequent characterization would provide additional information about incineration performance and might provide further assurance to the local population.

The schedule for the first 24 months of TOCDF operations is dominated by required trial burns. At least 16 trial burns (each consisting of multiple runs) are scheduled for the first two years. During these trial burns, samples will be collected and analyzed for products of incomplete combustion (PICs), particulates, heavy metals, and volatile and semivolatile PICs, such as chlorinated dioxins and furans. Data from the trial burns will be used to modify the health risk assessment conducted for the TOCDF. This information and the revised health risk assessment will then be used to develop an enhanced monitoring program for compounds identified as potentially significant. This is appropriate for the TOCDF, even though JACADS exhaust analysis indicates that the level of products of incomplete combustion will be very low. The Army appears to be committed to reassessing the sampling plan after conducting a health risk analysis of the data from the TOCDF trial burns.

Specific Recommendations for Laboratory Operations

The remaining specific recommendations addressed laboratory operations. The first recommendation of this group was the seventh specific recommendation:

> Increase the automation of sample handling and laboratory operations to ensure better quality control and efficiency. (MON-12)

The JACADS laboratory operations involve extensive handling of samples and manual data entry. Implementation of suitable and commercially available laboratory automation procedures could significantly enhance the efficiency and reliability of the laboratory.

Observers of laboratory operations at JACADS raised concerns that the process for handling DAAMS tubes appeared to be unnecessarily prone to error. Several steps have been taken at the TOCDF to improve these procedures. First, although desorption from the 8-mm DAAMS tubes to 4-mm tubes used for GC/FPD analyses is still required, a heated desorber has been developed and tested to mechanize this process. Ten tubes at a time can be desorbed with greater consistency than was previously possible. Second, much better control is exercised over the DAAMS tubes themselves. They are now individually barcoded, and a database is maintained on all uses and analyses by tube number. This should reduce errors of misidentified tubes, and allow tracking of tubes for gradual changes in agent retention or other deterioration of performance.

The eighth specific recommendation was:

> Give laboratory personnel a variety of tasks that ensure optimal attention and performance. (MON-13)

Repetitive tasks tend to decrease an operator's ability to detect unusual analysis results. An effective way to keep operators alert would be to include daily analysis of products of incomplete combustion in exhaust stack emissions.

Laboratory personnel can be assigned a variety of tasks to ensure that they remain alert and to provide breaks from repetitive tasks, particularly DAAMS tube analyses. One option is adding analysis of non-agent stack emissions. This analysis is now performed at the TOCDF, but the new task is not being used as part of a regular within-shift rotation for DAAMS analysts. A necessary condition for job enrichment is cross-training, which has not been completed in the laboratory operations because of the time pressures during systemization. There is some movement of operators between jobs on a longer-term basis, but this does not meet the requirement for varying tasks within a given shift.

The operators themselves were recruited from the local area, as well as from other chemical demilitarization and chemical weapons programs (e.g., CAMDS, JACADS, and Dugway Proving Ground, Utah). They consider their training at the Chemical Demilitarization Training Facility as good preparation for their jobs at the TOCDF because there are only minor differences between the equipment and procedures at the training facility and the disposal facility.

The ninth specific recommendation was:

> Give blind challenges to the laboratory. (MON-14)

The previous quality control procedures involved regular agent challenges to the laboratory chromatographs. These are more a calibration exercise than a verification that agent will be detected. Additional, unexpected challenges were recommended.

The TOCDF laboratory program, run by quality control personnel, includes randomly spiking DAAMS tubes with agent. Each challenge is chosen from a wide range of levels to prevent complacency among operators. Each operator receives at least one challenge per day. Both operators and their supervisors are blind to which tube is a challenge and to the level of the challenge.

The tenth specific recommendation was:

> Perform a detailed error analysis of the laboratory system and procedures. (MON-15)

Reliance on manual data entry and sample handling in the laboratory is a source of potential errors. A detailed error frequency analysis may reveal the sources and reduce the rate of errors in laboratory operations.

Reports of high error rates observed in JACADS laboratory operations prompted the committee to recommend an analysis to determine possible human errors and their consequences. At the TOCDF, a system hazard analysis has been conducted and presented in an Army report to the Utah DEQ as part of the RCRA permitting process, but this does not appear to cover errors in laboratory activities. Although there is evidence of a satisfactory response to detected errors, there is still a need for a system-oriented approach concentrated on potential errors. The Army contracted with MITRE to perform a laboratory error analysis during a five month period from May to September 1995. This analysis was scheduled to begin at JACADS and extend to laboratory operations at TOCDF. If performed adequately, this analysis should answer the Stockpile Committee's concerns.

Summary of Responses to Monitoring Recommendations

The Army has made considerable progress in responding to both the general and specific recommendations presented in the *Monitoring* report. Recommendations that could be addressed wholly or in part with existing commercially available instrumentation have generally been effectively addressed and implemented, with most solutions tested at JACADS prior to adoption at the TOCDF. The response to longer-range recommendations that require monitoring technology R&D has been more tentative. However, the monitoring improvements in the longer-range recommendations are not required for successful and safe operation at the TOCDF or other U.S. sites. They would, however, make operations easier, more efficient, and more reassuring to the public.

The Army and Tooele County now have a direct line of communication for informing civilian emergency management personnel of any alarms or alerts concerning agent release or other emergency incidents. An isolated false positive alarm requires stopping agent feed operations on-site but does not require activation of emergency response operations. A release large enough to threaten the immediate response zone (IRZ) would be quickly obvious on-site. A large release is likely to trigger several ACAMS alarms or to trigger the same ACAMS alarm several times. "Multiple" alarms are not characteristic of the response to false positives, which trigger sporadic, single alarms. Therefore, the response appropriate for a major agent release should not be triggered by sporadic ACAMS false positive alarms. At other sites, where the disposal facility is in closer proximity to communities, the emergency response may have to be activated sooner following an initial alarm. Thus, the false positive problem may have to be solved prior to the start of agent operations at those sites.

Committee visits to the TOCDF have included thorough discussions with laboratory and other monitoring personnel. Their training appears to be effective, and the equipment and operating procedures include significant improvements over those used during JACADS operations. The monitoring and supporting analytical laboratory capabilities in place at the TOCDF can support safe operations.

RECOMMENDATION ON CARBON FILTRATION

After examining the treatment of stack gases emanating from the baseline incineration system, the Stockpile Committee, in *Recommendations for the Disposal of Chemical Agent and Munitions* (NRC, 1994c) found that:

> The Stockpile Committee finds the baseline system to be adequate for disposal of the stockpile. Addition of activated carbon filter beds to treat all exhaust gases would add further protection against agent and trace organic emissions, even in the unlikely event of a substantial system upset. If the beds are designed with sufficient capacity to absorb the largest amount of agent that might be released during processing, addition of these beds could provide further protection against inadvertent release of agent.

Consequently, the committee made the following recommendation:

> The application of activated charcoal filter beds to the discharge from baseline system incinerators should be evaluated in detail, including estimations of the magnitude and consequences of upsets, and site-specific estimates of benefits and risks. If warranted, in terms of site-specific advantages, such equipment should be installed. (REC-13)

The principal focus of this recommendation was to consider whether charcoal filters might provide an additional safety factor at continental U.S. sites with relatively large nearby populations. Although the filters might reduce some nonagent emissions and could provide additional protection against plant upsets, the filters might also create additional risks if they caught fire, for example. The Army has initiated a review of various design options and related systems performance evaluations. If the evaluations are positive, the Army has chosen the TOCDF as the site for a demonstration unit; a primary reason for the selection of this site was the availability of an exhaust gas stream for testing purposes. This choice does not appear to be a prejudgment on whether a full-scale application of a carbon filter system would be warranted at the TOCDF or elsewhere.

3

Evaluation of Systemization Safety Performance

This chapter describes the direct and third-party information on which the Stockpile Committee based the assessment of safety performance during the Tooele Chemical Agent Disposal Facility (TOCDF) systemization (operational testing prior to the start of agent operations). The committee's evaluation is based on extensive reviews of safety-related audits, reports or studies made by other organizations on the TOCDF, on data and information requested and received from the Army, and on four visits of the Stockpile Committee to the TOCDF. In addition, specialist subgroups of the committee have concentrated on particular areas of concern by making additional visits to the TOCDF, doing spot-checks on site, and holding meetings with a wide range of personnel at the facility.

Next, Army procedures for a Pre-Operational Survey are discussed. The Army will use the Pre-Operational Survey as the basis for deciding whether to start agent operations at the TOCDF. Since this report predates the completion of the Pre-Operational Survey, the Stockpile Committee's evaluation will be limited to review of the plans and procedures that will be followed by the Army in completing this final review.

SAFETY-RELATED FUNCTIONS AND REVIEWS BY OTHERS

Systems Hazard Analysis

As part of the Army's safety program plan for the TOCDF, the Ralph M. Parsons Company performed a systems hazard analysis (SHA) to identify, evaluate, and quantify hazards associated with the activities within the TOCDF munitions demilitarization building and activities required to support the demilitarization process operations (Parsons, 1993). In simple terms, an SHA is a comprehensive and systematic search for and evaluation of all significant failure modes that can be identified by an experienced hazards analysis team. The SHA follows design drawings on a component-by-component basis for all systems and, for the TOCDF, utilizes additional performance information described in detail in the TOCDF Functional Analysis Workbook (U.S. Army, 1993–1995).

The SHA used various hazard and failure mode assessment techniques described in MIL-STD-882B. These techniques included failure modes and effects analysis (FMEA), fault tree analysis (FTA), and event tree analysis (ETA), used as appropriate (or in combination) for particular components or subsystems. Failure modes and effects analysis techniques are applicable to components where only single mode failures need to be considered. Systems or components with potential for multiple/sequential failures are better analyzed using FTA or ETA techniques. Twenty-one plant subsystems were analyzed:

- by FTA—electrical distribution and power system; compressed air supply; heating, ventilation, and air conditioning (HVAC); fire dampers; fire protection system; Automatic Continuous Air Monitoring System (ACAMS); central decontamination supply system; instrumentation and control system; hydraulic power system; brine reduction area and its pollution abatement system
- by FMEA—fuel gas supply; agent collection system (ACS); munitions handling; door monitoring; container handling building; dunnage furnace; dunnage pollution abatement system; liquid incinerator; metal parts furnace; deactivation furnace system; "wet" pollution abatement system

After the subsystem analyses were completed, two additional analyses using a combination of ETA and FTA techniques were performed to look for potential critical interactions between subsystems. The two interaction studies were focused on the following areas:

- HVAC, fire dampers, and fire protection
- agent and munitions handling

TABLE 3-1 Severity Ranking Criteria

Severity Level		Agent Release In-Plant	Personnel	System Loss
I	Castastrophic	IDLH[a] outside engineering controls	Illness, death, or injury involving permanent total disability	>25% and/or >1 month to repair
II	Critical	Greater than or equals PELs[b]/ASCs[c] outside of engineering controls or in ventilation area	Injury involving permanent partial disability	10% to 25%; 1 week to 1 month to repair in ventilation area
III	Marginal	Greater than or equals PELs inside nonagent areas	Injury involving temporary total disability	<10%; <1 week to repair
IV	Negligible	Less than PELs inside nonagent areas	Injury involving only aid or minor supportive treatment	No system loss, downtime, or repairs completed within 1 day

[a]IDLH = immediately dangerous to life and health.
[b]PEL = permissible exposure limit.
[c]ASC = allowable stack concentration.

Source: Parsons, 1993.

The SHA teams used standard procedures to identify potential failure modes and used failure rates from extensive databases established by the Department of Energy, the nuclear utilities, and the Institute of Electrical and Electronic Engineers. Components were assumed to be tested and serviced periodically and to be in an "as new" condition every 12 months. The plant operational life was assumed to be five years. Only failures during normal operation were considered; failures caused by external factors (earthquakes, severe flooding, tornados, fires, or other natural disasters) were not included in the SHA.[1]

In the SHA, each failure mode for each item was documented and discussed. Basic failure-initiating events were identified from models, considering root causes such as component failure, testing-initiated causes, and maintenance-initiated causes. If the failure could lead to (a) an agent release in the plant; (b) personnel illness, death, or injury; or (c) major system downtime, it was assigned a *severity* rating. Toxic vapor dispersion models (developed by the Chemical Research and Engineering Center of the U.S. Army Armament Munitions Chemical Command) were used to estimate downwind hazard zones from agent releases. Table 3-1 describes the severity ranking criteria used. Once the severity rating was established, the frequency of occurrence was estimated and frequency levels (events per year) were estimated. Table 3-2 shows the criteria used to establish qualitative frequency categories.

Finally, severity and frequency measures were combined to give a single indicator, the risk assessment code (RAC). RAC criteria are shown in table 3-3 (Parsons, 1993). As a result of the SHA activity, four potential RAC 1 events were identified and nine potential RAC 2 events. Recommendations were made for mitigating the effects of each event. Twenty-one potential RAC 3 events were identified, and controls were recommended for each.

Utah Department of Environmental Quality
Required Report for the Systems Hazard Analysis

The *Required Report for the Systems Hazard Analysis,* an Army report summarizing each of the potential RAC 1, 2, and 3 events, the SHA team recommendations,

---

[1]These failure modes are included in the risk assessment described in chapter 6.

TABLE 3-2 Criteria Used to Establish Qualitative Frequency Categories

Qualitative Frequency[a]	Agent Release In-Plant	Personnel Injury/Illness	System Loss
A — frequent	A 1E-01	A ± 10.0	A ± 1.0
B — probable	1E-01 > B ± 1E-02	10.0 > B ± 1.0	1.0 > B ± 1E-02
C — occasional	1E-02 > C ± 1E-03	1.0 > C ± 1E-02	1E-02 > C ± 1E-03
D — remote	1E-03 > D ± 1E-04	1E-02 > D ± 1E-04	1E-03 > D ± 1E-04
E — improbable	1E-04 > E ± 1E-06	1E-04 > E ± 1E-06	1E-04 > E ± 1E-06
F — not credible	1E-06 > F	1E-06 > F	1E-06 > F

[a] Events per year shown in exponential notation (e.g., 1E-02 = 0.01 events per year).

Source: Parsons, 1993.

and the Army's final mitigation actions, was submitted to the Department of Environmental Quality (DEQ) of the state of Utah as part of the TOCDF Resource Conservation and Recovery Act (RCRA) permit application (U.S. Army, 1993).

In a letter to the Army, dated August 11, 1994, the Utah DEQ transmitted approximately 70 comments and questions on the material submitted in the *Required Report*. On October 3, 1994, the Army submitted a comprehensive response to each of the issues raised by the Utah DEQ. At the time of the reply, some of the action items had not been completed. Most have been completed in the interim, but the Utah DEQ will require full compliance with the stated actions before issuing approval to start agent operations.

Facility Construction Certification

The RCRA permit for the TOCDF requires that the facility be constructed in accordance with the permit and its attachments. The state of Utah has defined 19 systems with critical components of the facility that must be certified to meet permit requirements. Four of these critical parts are tank systems. Certification of these tank systems was done according to RCRA procedures, namely, 40 CFR 264.192b–f. Because RCRA does not include procedures for certifying nontank systems, procedures for certifying the remaining critical parts of the facility were developed by the external engineering contractor, Forsgren Associates of Salt Lake City.

Systems Requiring Construction Certification

The critical systems requiring certification included seven waste-handling systems, some of which are hazardous waste management units (HWMUs) and some of which are supporting systems (that are not HWMUs). The following HWMUs required certification: the brine reduction area (BRA); liquid incinerator #1 (LIC); liquid incinerator #2; dunnage furnace (DUN); deactivation furnace system (DFS); metal parts furnace (MPF); and container handling building (CHB).

The pollution abatement system (PAS) and electrical interconnect system for the brine reduction area (HWMU 1) and all furnaces (HWMUs 2–6) are included as part of the HWMU. Each liquid incinerator includes the slag removal system. The tank systems that required certification are part of the brine reduction area; the agent collection system; the spent decontamination system; and the agent quantification system (AQS).

The state of Utah has required that eight systems that are *not* HWMUs also be certified. These non-HWMU systems are: electrical alarms, interlocks, and interconnect systems for the four furnaces and the brine reduction area and their associated pollution abatement systems; the halon fire suppression system; the heating, ventilation, and air conditioning filters; the air locks; the

TABLE 3-3 Risk Assessment Code (RAC)

Qualitative Frequency	Severity Level			
	I (Catastrophic)	II (Critical)	III (Marginal)	IV (Negligible)
A — Frequent	1	1	1	3
B — Probable	1	1	2	3
C — Occasional	1	2	3	4
D — Remote	2	2	3	4
E — Improbable	3	3	3	4
F — Not credible	4	4	4	4

RAC 1 — Hazards are unacceptable and measures must be taken to reduce the hazard to an acceptable level. If interim mitigation measures are used, approval to proceed must be granted by the top level designated in the Army's System Safety Management Plan.

RAC 2 — Hazards are undesirable and measures must be taken to reduce the hazards to an acceptable level. Interim mitigation measures require approval by the second-level authority designated in the Army's System Safety Management Plan.

RAC 3 — Hazards are considered acceptable with measures or controls taken to reduce the risk. Controls may be instrumentation, administrative, and procedural.

RAC 4 — Hazards are considered acceptable.

Source: Parsons, 1993.

ACAMS, the Depot Area Air Monitoring System (DAAMS), and the continuous emission monitoring system (CEMS); demilitarization machine systems; the deactivation furnace system blast enclosure; and the spare demister. Certification assumptions and procedures were the same for nontank HWMU and non-HWMU systems.

Construction Certification Assumptions

Because the permit was issued when drawings and specifications for the TOCDF were only 60 percent complete, a key portion of the certification procedure was review of the documentation through which the complete design was incorporated into the permit. The following assumptions were used for construction certification:

1. The purpose of the construction certification was to determine if certain systems at the TOCDF had been *constructed* in accordance with the permit. Conformity of the design and operation of the TOCDF with permit requirements was not part of this certification. No design calculations were verified.

2. Construction certification has been performed only for the 19 systems identified above and only for items within the defined boundaries of each system.

3. No equipment was dismantled as part of this certification. All hidden, buried, covered, or otherwise inaccessible items were evaluated based on a review of written records.

Document Review

The first step of the certification process was identification and review of facility drawings, specifications, and construction documentation. The purpose of this review was: (1) to identify the items in each system that required physical inspection; and (2) to identify and evaluate the documentation for items that could not be physically verified.

The following documents were used in information gathering for the certification: RCRA permit and attachments; redline (as-built) drawings; specifications; requests for information; engineering change proposals; turnover packages; and construction inspection reports.

The information described in the documents listed above was used to determine which items in each system required certification. After the documentation was reviewed and the HWMU or system being certified was visually inspected, a Certification Report was prepared summarizing the conclusions and identifying deficiencies. The following sections describe the certification procedures.

System Description. The first step in the certification process was to accurately define the system or HWMU being reviewed. A description was prepared establishing the boundaries of each system or HWMU, including equipment and intended use.

Documentation Review. Based on the descriptions prepared for each system, the appropriate documentation (e.g., specifications, engineering change proposals, vendors' information packages) was requested. The information was reviewed and specific areas lacking appropriate documentation were noted.

Inspection Checklists. Based on the document review, checklists were prepared of items that required visual inspection. In most cases, meeting specifications satisfied the requirements of the checklist. When installation instructions from equipment manufacturers were available, they were used as guides for the field inspections.

Field Inspection. Each system was observed in the field using redline drawings and checklists to structure the inspections. Checklists were often expanded or modified based on visual observations or field conditions. All completed checklists, as modified for each system, are included in the Certification Report as attachments.

Certification Report. After the field inspection and review of additional documentation, a Facility Construction Certification (FCC) report was prepared for each system or HWMU that was inspected. The report summarizes the findings, identifies nonconformance with the permit requirements, describes the certification procedures, and contains a certification statement signed by a professional engineer (Forsgren Associates) registered in the state of Utah.

The FCC report was then reviewed with Edgerton, Germerhausen and Grier, Inc. (EG&G, the Army's systems contractor), to identify all nonconformance and unresolved issues and to develop a "work list" for analysis. The outcome of the FCC report review was either (1) permission by the Division of Solid and Hazardous Waste (DSHW) to proceed with systemization; or (2) development by DSHW of a "punch list" of unresolved issues the TOCDF must address to satisfy FCC requirements. Once the FCC reports have been approved by DSHW, the TOCDF can proceed to operational readiness/systemization using hazardous materials. The Facility Construction Certification process as outlined in figure 3-1 ensures that the TOCDF has demonstrated that all systems are designed and constructed to protect human health and the environment. All 19 critical systems require FCC approval prior to the initiation of agent destruction.

Inspector General Report
Courtesy Chemical Surety Inspection—Tooele CDF

At the request of the office of the Program Manager for Chemical Demilitarization (PMCD), inspectors from the U.S. Army Inspector General Agency conducted a courtesy chemical surety inspection of the TOCDF during August 15–18, 1994. The team performed the inspection as if the facility were in operation, although systemization testing was still under way, many procedures were still in draft form, and some final construction was still in progress. The results of the inspection were summarized in a report (U.S. Army, 1994a) transmitted to the Commander, U.S. Army Chemical Materiel Destruction Agency (now Program Manager for Chemical Demilitarization), on September 6, 1994. Fifteen pages of text described deficiencies found during the inspection. Although this may seem like a long list, it is, in fact, a short list for a facility of this size and complexity. The following topics were covered: mission operations; safety; security; surety management; accident/incident response and assistance; and external support. Many of the deficiencies

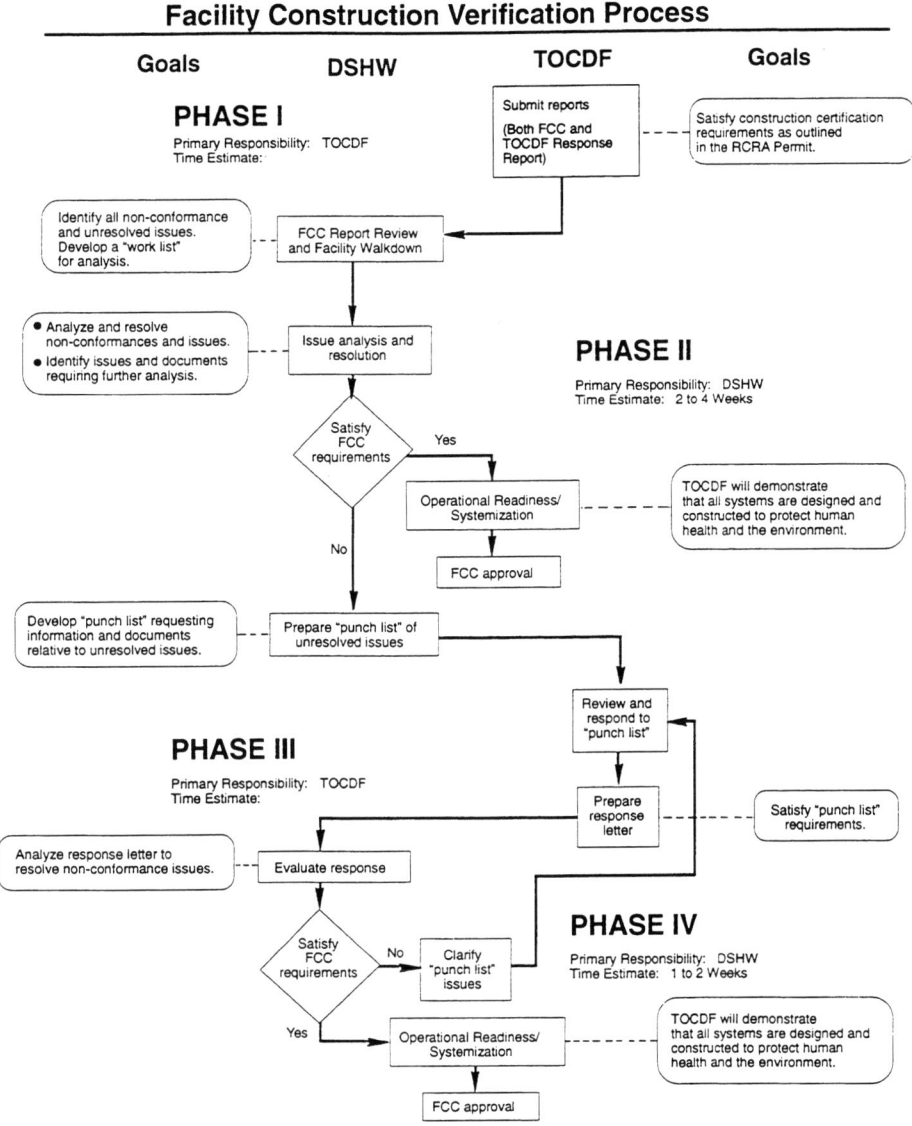

FIGURE 3-1 Outline of the Facility Construction Certification Process. Source: EG&G, 1995b.

involved documentation and procedural matters, incomplete training, problems in the medical clinic programs, and concerns about staffing at both the TOCDF and the Utah DEQ for facilitating approval of necessary permits.

On September 29, 1994, the Army submitted a memorandum containing detailed responses to each of the items raised in the Inspector General Courtesy Chemical Surety Inspection report. This memorandum discussed time schedules and corrective actions for procedures and programs that were found deficient, and it concluded that the facility would be in compliance before the start of agent operations.

TOCDF Safety Evaluation Report

In September 1994, Mr. Stephen Jones, who had recently been terminated by EG&G Defense Materials, Inc., from his position as the TOCDF safety and security manager, put forth allegations of 119 safety-related deficiencies at the TOCDF. The Army Safety Office was directed to perform an investigation into each allegation raised by Mr. Jones. The Army Safety Office completed a *Safety Evaluation Report* (U.S. Army, 1994b) on these issues and the Secretary of the Army released the report to Congress. The 166-page document cites each of Mr. Jones's claims and the report of the person assigned to do the investigation. Approximately half of the claims were found to have some substance, but many of the deficiencies were typical of a plant not yet in operation. A critical reading of the report revealed some inattention on the part of TOCDF personnel to detailed procedures, as well as some problems with the maintenance of auxiliary equipment (e.g., hoist cables, forklifts). The Army Safety Office concluded that the current safety and health risks at the TOCDF are "minimal and acceptable." The Army has indicated that individual allegations that were found valid by the Army Safety Office either have been corrected or will be corrected prior to the start of agent operations.

U.S. Army Chief of Engineers Report
TOCDF Report on Design-Related Safety Issues and Evaluation of Construction Conformance with Design

The Assistant Secretary of the Army for Installations, Logistics and Environment requested that the Army Chief of Engineers review the TOCDF Safety Report to determine if any of the 119 safety allegations by Mr. Jones had design implications. Further, the Chief of Engineers was asked to confirm that the plant had been constructed in accordance with the design.

A nine-member team was formed to accomplish this task (U.S. Army, 1994c). Of the 119 allegations addressed in the TOCDF safety report, only 12 were found to be "design-related." None of the 12 was found to compromise either the safety or the environmental integrity of the plant. All were essentially enhancements to improve worker safety during operations. Six of the items reviewed had already been corrected by the time the report was issued, and the remaining six had been scheduled for prompt resolution.

With respect to conformance to design, the team reviewed the ongoing Facility Construction Certification Process, required by the State of Utah for RCRA permits. The summary conclusions of the Chief of Engineers were:

- The Facility Construction Certification Process is consistent and thorough. The consultant's certification reports maintain a consistent format that is clear and concise. Outside references are kept to a minimum and the reports are very detailed. Assumptions were clearly identified in each report. In summary, these reports are effective, stand-alone documents which clearly evaluate and identify the extent to which construction complies with the RCRA permit.
- The team firmly believes that another consultant operating under the same procedures and methodology would essentially develop the same findings and conclusions. We believe that the repeatability of the certification results is very high.
- In summary, (1) none of the allegations indicate a flawed or inadequate design; and (2) there is a comprehensive, rigorous, professional program in place that ensures that construction conforms to the permitted design.

U.S. Army "Lessons Learned" Programs

Over the past several years, the Army has implemented a Programmatic Lessons Learned (PLL) program (U.S. Army, 1995c). The objective of the program is stated as follows:

> The PLL program consists of an organized and managed system which facilitates the extraction and conveyance of information (Lessons Learned) from and to the primary program focal points. PLL is a method for (1) identifying both effective and ineffective plans, policies, procedures, or actions; (2) providing recommendations and actions to strengthen the chemical demilitarization program; and (3) causing tangible changes to the overall program (changes to equipment, contracts, plans, policies, procedures, etc.).

The PLL program operates under a PLL Control Board, consisting of a chairman (Chief of Operations Division of the Office of the Program Manager for Chemical Demilitarization), board members representing U.S. Army Chemical Demilitarization and Remediation Activity (now Program Manager for Chemical Demilitarization) organizations, and support staff. The PLL Control Board reports to the Program Manager for Chemical Demilitarization but has authority for funding and the programmatic implementation of approved PLLs, up to $750,000 per directed change.

In its first two years, the PLL program concentrated on gathering information (from all sites, organizations, and program phases), and documenting, evaluating, and identifying actions to be taken. Each identified action was assigned to an organization, which then became responsible for implementation. In early 1995, the PLL program was expanded to include subject area workshops, bringing together

Army and contractor management and operational personnel from active sites to solicit lessons learned from multiple experts in 19 focused subject areas. The workshops included briefings and roundtable discussions. Any changes/actions are referred to the PLL Control Board for approval and implementation, if appropriate. The subjects of workshops conducted to date include: incinerator operations and maintenance; pollution abatement systems; the brine reduction area; the rocket handling system; the demilitarization process lines and associated training; general operating issues; plant management; laboratory operations; monitoring operations; and environmental issues. Prior to the scheduled start of agent operations, additional workshops are planned. These include: protective clothing; emergency response; medical support; depot issues; support systems; spare parts; construction; systemization; and a second workshop on incinerator operations and maintenance, pollution abatement systems, and the brine reduction area.

The PLL program provides a forum for people from different sites to share experiences so that improvements can be implemented in a consistent manner. Because there are different contractors and personnel at each site, the PLL program is a useful mechanism for identifying and managing opportunities for continual improvement throughout the program and maintaining consistency in practices from one operating or training site to another.

In addition to the PLL program, the Army has also established a Field Lessons Learned Review Team (FLLRT). The purpose of the FLLRT is to provide central documentation and review for all field-approved engineering change proposals, design-related requests for information, or documented lessons learned at demilitarization facilities currently under construction, systemization, or operation and to provide direction for the preparation of consistent engineering change proposals for other sites. If the FLLRT disagrees with changes submitted for review, the team is required to offer comments to the initiating site Configuration Control Board and to follow up on resolving the disagreement.

The permanent members of the FLLRT include the chairman (from the Chemical Stockpile Disposal Program Design Branch) and representatives from the Operations and Training Branch, the Equipment Acquisition Branch, the Safety Branch, the Quality Assurance Branch, and the Environmental and Monitoring Division. The FLLRT meets monthly, and the databases are maintained by the Army's design and systems integration contractor. The Army is also preparing an assessment of the effectiveness of both individual and overall lessons-learned programs. These evaluations will be performed by independent assessment teams.

U.S. Army Subject Area Review Reports

The office of the Project Manager for Chemical Stockpile Disposal (PM-CSD) has established a Subject Area Review (SAR) program. In this program, particular subject areas are identified (e.g., management, documentation, unique operations) that are important to the safe and efficient operation of the TOCDF, as well as to ensuring compliance with regulatory requirements. The PM-CSD office appoints a review team of knowledgeable and experienced people within the Chemical Stockpile Disposal Program (CSDP). The team develops a review plan, spends several days at the site, and prepares detailed comments and recommendations. (The Stockpile Committee requested, received, and reviewed selected subject area review reports).

For example, a five-person review team prepared a subject area review report called "Major Program Documentation." Safety, operations, maintenance, laboratory, and administrative procedures at the TOCDF were evaluated—some by review of documents, some by selected "walk-downs" of procedures (going through a procedure item by item to determine completeness and adequacy), and some by discussion with responsible staff. The report contains 20 recommendations, some of which include detailed suggestions for a number of improvements in the area under review. Because the review was conducted in February 1995, some documentation was still incomplete or not fully reviewed. However, the review team noted these deficiencies and made recommendations for completing action items prior to the end of systemization. For example, they recommended comprehensive walk-downs of preventive maintenance procedures to determine if changes or improvements were needed. They also established requirements for documenting the completion and review of all major program documentation. Major revisions were recommended for some laboratory procedures and the quality control plan. Meetings were scheduled to review progress on all the recommendations, so that all major documentation would be in place and reviewed prior to the end of systemization.

Another subject area review team examined management practices during a site visit to the TOCDF in

February 1995. Suggestions were made for improving the TOCDF Management Plan, for establishing specific inspection/surveillance requirements for meeting each contractual obligation, and for finalizing the award fee criteria by June 1, 1995 (the deadline has since been changed). The review team also recommended that the office of the Program Manager for Chemical Demilitarization develop consistent policy statements or guidelines to aid the facilities in their interaction with outside groups, including state agencies, other federal agencies, citizens advisory commissions, and the news media. At present, outside inquiries are being handled at the TOCDF on an individual basis by the TOCDF project manager. Suggestions were also made for improving procedures and training for shift supervisors to cover issues of oversight and communication during shift changeovers. Finally, the review team recommended the development of a "critical activities" manual to identify activities important to safety and/or environmental compliance. The Johnston Atoll Chemical Agent Disposal System (JACADS) critical activities manual will be used as a basis for developing an appropriate manual for TOCDF.

As of this writing, 13 subject area reviews have been completed: Emergency Response; Network Analysis System; Major Program Documentation; Training and Qualifications; Surety Program Quality Assurance/Quality Control; Management/Oversight; TOCDF Unique Operations; JACADS Lessons Learned; Medical; Administration; Depot Support; Chemical Stockpile Emergency Preparedness Program (CSEPP) Interface; and Physical Security. The subject area review program has been effective so far in identifying needed improvements in TOCDF systems, procedures, and management and has established schedules and check points for implementing recommendations of the review teams.

The subject area review team generates a draft report with recommendations. That draft report is provided to the TOCDF project manager, to the systems contractor (EG&G), and other personnel involved in the TOCDF project for review and comment. The subject area review team reviews these comments and incorporates the ones they agree with. Other comments are discussed with the proponent. If no resolution is reached, the disagreement is referred to the PM-CSD for resolution and direction.

Once the comments have been incorporated, the report is submitted to the PM-CSD for approval. This approval is sometimes accompanied by additional directions based upon the PM-CSD review of the Subject Area Review report. Upon signature of the PM-CSD, the subject area review report becomes directional to the TOCDF project manager and to other elements in the office of the Program Manager for Chemical Demilitarization. This means that the Program Manager must approve any deviation from recommended actions.

As part of the operational readiness evaluation and Pre-Operational Survey process, the status of subject area review actions will be reviewed to ensure their proper completion.

State of Utah Inspections

The Utah Department of Environmental Quality (DEQ) has established a number of requirements as part of the RCRA permitting process. In addition to the *Required Report for the Operational Verification Tests* (U.S. Army, 1993) and the Facility Configuration Certification described earlier in this section, Utah DEQ inspectors are on site at the TOCDF several times a week during the systemization period. As engineering changes are proposed by the site contractor and approved by the Army, they are followed closely by the inspectors to facilitate processing changes that might require modification of the permit.

The Utah Industrial Commission, Occupational Safety and Health Division, has also conducted an on-site inspection to evaluate worker safety (in addition to three inspections by the U.S. Department of Labor).

STOCKPILE COMMITTEE SITE VISITS

The Stockpile Committee visited the Tooele Chemical Agent Disposal Facility during construction (November 1991 and March 1993) and during systemization (May 1994 and March 1995). Since February 1995, four subgroups of the Stockpile Committee have visited the site to make more detailed evaluations of particular portions of the facility and planned operations. The first subgroup visited on February 23–24, 1995, with the objectives of examining changes in major equipment as a result of recommendations in the OVT report, Part II (NRC, 1994a). The subgroup concentrated on systems for engineering change control, safety and environmental management, and relevant documentation control (e.g., procedures, compliance documentation); and management structures for assuring safe, environmentally sound operations. A second subgroup visited the facility on March 28, 1995. This group focused on operations, maintenance, automated control systems, and management of change, spare parts, and training. A third subgroup visited on March 29, 1995, and focused on laboratory operations and monitoring. Finally, a fourth subgroup visited the TOCDF on June 15, 1995, to review safety management issues.

The changes made to the baseline system at the TOCDF in response to recommendations by the Stockpile Committee are described in chapter 2. Key observations by the subgroups that pertain to systemization performance follow.

Personnel Issues (Recruitment, Training, Turnover)

Competence at all levels is a prerequisite to safe operations in any organization. The committee was informed by the TOCDF general manager that his plan to recruit and retain highly qualified employees was based on his organization being a "premier employer" that offers good training, attractive compensation, and a good managerial environment. The following observations concerning competence are based on interviews with first-level operators, such as control room operators, in-plant operators, maintenance technicians, and laboratory technicians. Most of these employees were recruited from the local labor pool (Tooele County and surroundings) and from other Army chemical operations (JACADS, CAMDS, and the Dugway Proving Ground, Utah). The operators interviewed by the committee gave evidence of relevant backgrounds, intelligence, and the desire to take advantage of the opportunities their jobs offered, such as studying at local colleges.

Operators are trained both at the Chemical Demilitarization Training Facility (CDTF) at Aberdeen Proving Ground, Maryland, and at the TOCDF. When questioned, operators volunteered detailed explanations of procedures, described improvements that had been made during systemization, diagnosed potential upsets, and demonstrated knowledge of specific procedures for dealing with them. The training philosophy of emphasizing knowledge-based understanding as a basis for rule-based procedures appears to have been successfully assimilated by the operators.

Upset drills cannot be performed with the site control panel computers, which are reserved for operational control. As of this writing, it is not clear if operators will return to CDTF periodically for upset training. The Army's management approach is to do all training on site; on-the-job-training at the panels is believed to be the best training, with walk-through simulation for upset drills. The committee and some individuals in the EG&G organization and PMCD headquarters believe refresher training at the CDTF using consoles and process simulators is necessary. Although most upset conditions are believed to be straightforward, the committee believes that realistic upset drills at the consoles are needed for first-rate preparedness.

Another training principle, that of training all levels of plant personnel together who work on the same shift, appears to be producing better understanding between groups. Lack of this rapport had been identified as a shortcoming by the Stockpile Committee in early visits to JACADS, where operators and maintenance personnel worked under different subcontractors.

Cross-training is beginning at the TOCDF. Control room operators are typically certified on at least two systems. In addition, they often accompany in-plant operators for a better understanding of plant operations. Cross-training programs are expected to be expanded in the future, although cross-training has not yet been implemented in some areas, such as the laboratory. In maintenance, which is based on trade skills acquired before employment at the TOCDF, there is less scope for full cross-training between distinct specialties, between electrical technicians and welders, for example.

Training in process operations and agent operations appears to be thorough, but training in general safety practices requires improvement. Some deficiencies observed during the walk-through of demilitarization process areas include open electrical cabinets with exposed conductors with no maintenance personnel or physical barriers present, malfunctioning eye wash stations, lack of clear and enforced practices for protective clothing, conflicting signs for ear and eye protection, and tools left in walkways.

A General Observation

There appears to be a general belief at the TOCDF that safety practices are primarily for agent operations. As a result, the emphasis on safety has been focused on agent-related issues with less emphasis being given to industrial safety practices. The committee believes this is not acceptable for the following reasons:

- It will be impossible to change habits suddenly when agent operations start.
- Accidents can happen now as well as when agent operations start.
- Although agent has yet to be introduced, the TOCDF is no longer a construction site. At this stage of systemization, it is an operational facility.
- Even for a construction site, better housekeeping and safety practices are appropriate.
- Plant industrial safety practices still require improvement.

Shift Operations

The work pattern is based on 12-hour shifts in a standard pattern, rotating between three per week or four per week, with seven-day breaks. This work schedule was considered an advantage by most operators once they had adapted to it. There is little information in the current human factors literature about the frequency of errors at the end of long shifts, but the literature on the subject should be followed closely as more is learned. Because shifts are so long, overtime added to the end of any given shift should be avoided. Overtime was running at a low level during systemization, although one employee claimed to have averaged more than 30 hours of overtime per month during a three-month period. The overtime system appears to be well controlled overall, but such high use, even by a few people, should be closely controlled.

The shift-change system was well designed and executed. During the half-hour overlap period (which accounts for some of the overtime), there are face-to-face briefings at all levels, not just between leads and supervisors. There is also a written handoff briefing, which was conscientiously used in a sample examined by the committee. Also, at the start of each shift, there are operator and supervisor meetings as well as plant-wide coordination meetings. Although poor shift-change procedures have led to accidents and incidents in other systems, such as aviation maintenance systems, this does not appear to be a problem at the TOCDF.

Maintenance and Spare Parts

Maintenance is divided broadly into preventive maintenance and on-demand maintenance. Preventive maintenance is assigned to a dedicated crew and is performed in accordance with a designated schedule indicating maintenance intervals, which can be daily, weekly, or monthly. Most of the maintenance is performed using written procedures and checklists. Increasingly, daily preventive maintenance is being assigned to operators as part of their standing operating procedures. This should improve the (already adequate) relationships between maintenance personnel and operators and provide operators with more detailed knowledge about the state of the equipment. The procedure for nonroutine repairs and maintenance begins with the issuing of a Work Request, which is assigned to a particular maintenance crew. Written procedures are followed for most tasks. The exception is for trade activities when it is not feasible to pre-specify every weld. Here, more general procedures are used by the skilled tradespeople, all of whom have been trained at the Chemical Demilitarization Training Facility. A Safe Work Permit is issued for each work request from the control room, so that the control console operators are aware of all maintenance activities and the areas where maintenance work may be in progress. The Safe Work Permit is also used to check and coordinate activities, such as lockout and tag procedures and equipment status verification, between area operators and maintenance personnel.

Spare parts are handled by the warehouse through a computer inventory system that specifies minimum and maximum stock levels and reorder procedures for each of 15,000 or so parts. Levels are initially based on the manufacturer's recommendations and prior experience at JACADS. Levels will be adjusted to reflect subsequent operating experience. Goods received at the warehouse undergo quality control checks, and none will be issued without the applicable Material Safety Data Sheet. Stock outages have not led to changes in the automatic reordering policy so far. The recurrence of expensive and time consuming errors in other industries makes the committee uncomfortable with this reactive approach. During operations, any shortage of parts could be critical and could lead to pressures for alternative, perhaps less safe, ways of maintaining operations. There are standard methods of inventory control in other industries that ensure very high levels of stock out protection without excessive inventory costs.

There are no formal controls at the TOCDF to account for tools left inside controlled access areas. TOCDF management indicated that other measures are sufficient to prevent this problem (e.g., workers have to pay for lost tools; operations will not sign off on work until the area is cleaned up).

General Management Issues

A safe plant requires strong, but open, leadership and an appropriate organizational structure. The site visit team was generally impressed with the quality and style of the top leadership at the TOCDF. The organizational structure is necessarily complex in a plant with multiple activities and contractors but did appear to be appropriate. The EG&G management structure was designed to mirror, to some extent, the Army leadership structure under the office of the Program Manager for Chemical Demilitarization, with parallel elements reaching to an extensive level deep within the organization. All relationships between the Army, contractors, and subcontractors appeared to be working well.

Within each shift there is a matrix organization that seems effective. There is a designated plant shift manager to whom all section managers report, whether from operations, maintenance, or laboratory (run by Battelle Corporation). No friction was observed between people within this system.

An important tool for any organization to evaluate operating procedures is to "benchmark" performance with similar organizations. Data were provided to the visiting team by Science Applications International Corporation (SAIC) on benchmarking safety procedures, but not on other aspects of operations, such as organizational design, laboratory work, or maintenance performance. SAIC presented data to the Stockpile Committee comparing the JACADS and TOCDF lost-time injury rates and recordable injury rates with the rates in comparable industries (SAIC, 1995c). Compared with the top 10 percent of chemical companies and with specific leading companies, rates of lost-time injuries were similar at JACADS and the TOCDF, but worse in terms of total recordable injuries. Although there are special circumstances that may affect the data (e.g., construction activity at the TOCDF and size of organization), the data still indicate a need to devise better indicators of safety than lost-time accident rates. Even recordable injuries are rare enough for rates to fluctuate too much for tracking purposes. Human factors engineers have used near-accidents and self-reported error incidents as more sensitive safety indicators in relatively safe plants. A system to capture information about precursor incidents should be established at the TOCDF and the other sites to provide early warnings of potential complacency. Indeed, as each site progresses from the construction phase through systemization to regular operations, care must be taken to maintain a high level of safety performance.

Finally, the TOCDF management should be commended for using a Chemical Personnel Reliability Program (CPRP), which establishes no-penalty checks on fitness for work and helps reduce the probability of sub-optimal performance at all levels.

Programmatic Issues

Certain issues transcend the boundary of the TOCDF and extend to the Chemical Stockpile Disposal Program itself. During the development of each site, new technology and procedures will be introduced that will not be immediately retrofitted at older sites. This has already happened between JACADS and the TOCDF. The Chemical Demilitarization Training Facility should reflect such changes. Ideally, the training facility could reconfigure equipment and procedures to match the exact needs of each cohort of trainees. The committee recognizes the difficulty of maintaining disparate system setups. At present, there is some lag time between the equipment at the TOCDF and the equipment at the training facility, although all operators interviewed found the training facility to be a good simulation of what they encountered at the TOCDF. The human factors literature on simulation fidelity needs to be reviewed to ensure that compromises in the direction of "generic" systems at the training facility have minimal impact on the transfer of knowledge and skills to actual operations.

A second aspect of systems configuration, configuration control, is also important. Permits and good engineering practice require careful consideration of the potential effects of changes in system configuration, e.g., changes in hardware, software, procedures, and training. The configuration control process now in place is designed to ensure that system safety is not compromised. However, this process should never be used to discourage innovations by operators. Operators can often see "better" ways of performing their tasks, but the lowest priority for change is often assigned to improving productivity, even though such changes may improve operator performance and reduce the possibility of human error. A two-way dialogue needs to be established and maintained to ensure that operators' suggestions are not just evaluated and dismissed but are taken as a starting point for process improvements. The TOCDF has instituted processes to move important procedural changes through the organization quickly. A number of working level people interviewed by the committee were not familiar with the process; other operators reported that changes they had suggested had been implemented. Thus, the management change process is working but needs continuous encouragement to remain effective.

The focus on agent operations as the trigger for implementing safety practices appears to be program-wide. The only way to instill a viable overall safety culture is from the top. Program management and site management must be committed to safety and must tolerate nothing less throughout the organization. More attention must be paid to nonagent safety issues to promote a total safety culture at the facility and to ensure a safety record at a level of the best comparable industry practice.

A final program-wide issue is the programmatic lessons learned initiative, described earlier in this chapter. Several people interviewed from the TOCDF and the office of the Program Manager for Chemical Demilitarization characterized this initiative as a highlight of the disposal program. The PLL should remain in place, as it will become even more important once agent operations start at the TOCDF.

PRE-OPERATIONAL SURVEY

Since 1973, the Army has required that all new or modified chemical disposal facilities undergo an extensive Pre-Operational Survey prior to the start of toxic operations. The Pre-Operational Survey is a formal review and assessment of facility operations for compliance with applicable safety, environmental, quality assurance, and surety standards. This review and assessment by the Army is to ensure that the TOCDF is capable of performing its functions in accordance with statutory and regulatory requirements and that an acceptable level of safety, environmental compliance, and operational performance can be achieved when agent operations start. The survey is conducted under the authority of the Program Manager for Chemical Demilitarization, and the survey team chairman acts as the Program Manager's designated representative. The team includes functional area team leaders, who assist in the technical and administrative management of the team, and individual members, who are considered experts in their fields. Table 3-4 is a listing of the team members.

In May 1995, a TOCDF Operational Readiness Evaluation was performed under the direction of the PM-CSD. The purpose of this evaluation was to assess the status of the TOCDF preparations for starting agent operations, with emphasis on program documentation and facility and equipment status. The evaluation team also referred to the results of earlier subject area reviews and assessed progress in completing the action items recommended in the earlier reviews.

The Pre-Operational Survey was scheduled for the fall of 1995 to review all pertinent documentation, including approval documents, directives, plans, procedures, and records; to conduct a detailed inspection of the facility; to perform an evaluation, under operating conditions, of all equipment and processes associated both directly and indirectly with the disposal of M55 GB rockets; to witness and critique selected system and subsystem tests; and to evaluate responses to a number of simulated contingency scenarios. The operational survey team will also witness and evaluate integrated plant operations under actual plant operating conditions during the processing of simulant M55 rockets. During the survey, all equipment, control systems, monitoring systems, and support systems and procedures will be exercised as if actual chemical munitions were being processed. All activities, starting with munitions handling and ending with treatment of demilitarization process residues, will be witnessed and evaluated by the survey team. Specifically, the areas of interest to the survey team will include:

- compliance with all statutory regulatory requirements, especially safety, environmental, occupational health, surety, and security requirements
- operational capability of process and support equipment (4-hour run with simulated rockets), facilities, procedures, and personnel; this includes storage yard, transportation, laboratory, laundry, maintenance, and supply functions
- process control, including design and function of hardware and software
- adequacy of program documents including plans, procedures checklists, and limiting conditions of operations;
- configuration management
- quality assurance/quality control

TABLE 3-4 TOCDF Pre-Operational Survey Team Members

Name	Affiliation
Chairman's Team	
Robert Perry	Chief, Risk Surety Management Division, Program Manager for Chemical Demilitarization (PMCD), Chairman
Clifford Dunseth	HQDA Safety Office
Dianna Rickets	Chemical Surety Officer, PMCD, Assistant
Carol Bieniek	Risk Surety Management Division, PMCD, Team Secretary
Quality Team (Quality Assurance, Training, Documentation, and Configuration Management)	
Tom Kartachak	Chief, Quality Assurance Branch, PMCD, Team Chief
Nick Stamatakis	Quality Assurance Branch, PMCD, Quality Assurance
Andrew Roach	Project Manager for Chemical Stockpile Disposal (PMCSD), Training
Morita Bruce	PMCSD, Configuration Management
Mike Pratt	Science Applications International Corporation (SAIC)
Surety Team (Surety, Security, Ammunition Surveillance, CSEPP, Emergency Response)	
Jeffrey Principe	Surety Officer, Chemical and Biological Defense Command (CBDCOM), Citizens Advisory Committee (CAC), Team Chief
Rick Knutson	Security Specialist, Tooele Army Depot (TEAD), Physical Security
Linda Hodgson	Security Specialist, Pine Bluff Arsenal (PBA), Physical Security
[Not available]	Surety Representative
Barbara Parsley	CBDCOM CAC, Emergency Response
Joe Miller	Chemical Stockpile Emergency Preparedness Program (CSEPP) Office, PMCD, CSEPP
Health/Environmental Team (Air Monitoring, Laboratory, Environmental, Medical)	
Dr. John Liddle	Department of Health & Human Services (DHHS), Team Chief
Michael Gooden	Environmental Monitoring Division, PMCD, Monitoring/Laboratory
Joseph Stang	Environmental Monitoring Division, PMCD, Environmental
Jay Smith	Department of Health & Human Services, Statistician
[Not available]	Representative, Tooele Army Depot Environmental
[Not available]	Representative(s), Raytheon Engineers & Constructors (RE&C), Environmental
Dr. Roger McIntosh	SAIC, Medical

continued

TABLE 3-4 Continued

Name	Affiliation
[Not available]	Representative, Headquarters Department of the Army (HDQA), Surgeon General's Office, Medical

Safety/Operations Team (Equipment, Facilities, Procedures, Worker and Public Safety)

Name	Affiliation
Gregory St. Pierre	Chief, Safety Branch, PMCD, Safety
C.T. Anderson	Safety Branch, PMCD, Safety
Bernard Bindel	Safety Branch, PMCD, Safety
Cheryl Maggio	PMCSD, Operations
Ralph Rogers	U.S. Army Center for Health Promotion & Preventive Medicine (USACHPPM), Industrial Hygiene
Harvey Rogers	DHHS, Safety Operations
Harold Oliver	TEAD Safety Office, Safety
Dale Druyor	PMCSD, Protective Clothing/Operations
Chuck Papish	PMCSD (JACADS), Operations
Buddy Webster	RE&C, Furnaces/Process Control
Steven Blurk	U.S. Army Technical Center for Explosives Safety (USATCES), Safety
Boyce Ross	Corps of Engineers, Huntsville Division (CEHND), Facility Design/Operators
Craig Adams	General Physics (GP), Control Systems
Sam Blouser	PMCSD (CAMDS), Operations
Tom Cain	MITRE, Munitions Tracking

Support Team (Storage Yard, Transport, Munitions Handling, Maintenance, Supply, Laundry)

Name	Affiliation
Sonny Smith	Anniston Army Depot (ANAD), Team Chief
Warren Taylor	PMCSD, Maintenance Supply
Craig Hatfield	PMCSD, Maintenance Supply
Raymond Cornier	PMCSD (CAMDS), Munitions Handling, Laundry
Tim Baker	PMCSD, Maintenance, Supply, Storage Yard
CW4 Jose Medina	CBDCOM Operations Directorate, Munitions Handling, Transport
Doug Maddox	CBDCOM CAC QASAS, Storage Yard, Transport, Munitions Handling, Laundry
Dave Underwood	GP, Maintenance

- agent and nonagent monitoring and detection
- handling, testing, inspection, and use of protective clothing and equipment
- safety systems, including process safety interlocks, alarms, lightning protection system, and ventilation system
- in-plant audible, visual, and written communications
- operator training and certification
- integration of chemical demilitarization activities, including support activities and emergency response activities
- overall control of people and materiel
- calibration and testing program
- recordkeeping and document control
- environmental compliance and solid waste management
- agent and munitions accountability

During the survey, findings are classified into three categories: I, II, and III. All Category I findings (essential to the safety of personnel or the operational readiness of the system) are required to be resolved prior to the start of agent operations. The Pre-Operational Survey team is required to verify in writing the closure of each Category I finding. Category II findings (items not considered as critical to the safety of personnel or the operational readiness of the system) require scheduled resolution. Category III findings (noted by a survey team and usually can be corrected on the spot) are suggested improvements. Once the Category I findings are closed out satisfactorily, the site can request permission from the Office of the Program Manager for Chemical Demilitarization to start agent operations.

The authority to start agent operations at the TOCDF is vested with the Program Manager for Chemical Demilitarization. This process begins with notification from the TOCDF project manager that the project team believes the facility and the work force are prepared to start agent operations. The Program Manager evaluates this recommendation along with the status of all Pre-Operational Survey findings. In addition, the status of all regulatory-related actions is reviewed to ensure that all regulatory requirements have been satisfied. Given the visibility of the demilitarization effort, the Army chain of command above the office of the Program Manager for Chemical Demilitarization will be fully notified of actions leading to a decision to start agent operations at the TOCDF.

DISPOSAL PROGRAM STAFFING

The PMCD staff has grown in order to perform the coordination and oversight functions required during the systemization of the TOCDF. This staff is able to manage effective oversight and interaction with the prime contractor at the TOCDF in areas relating to safety management, environmental management and compliance, design and construction certification, management of change and documentation, the "lessons learned" programs, the subject area reviews, and coordination with external agencies involved in health, safety, and environmental compliance, community relations, and emergency planning. Where necessary, detailed work has been delegated to experienced contractors. The *Recommendations* report stated concerns about disposal program staffing as follows:

> The Army should establish a program to incrementally hire (or assign military) personnel to ensure that staff growth is consistent with the workload and with technical and operational challenges. These additional personnel must be assigned and trained before the project office gets deeply involved in addressing each challenge. (REC-21)

The committee finds that the present PMCD staff is working very hard and appears to be doing a sound job managing the elements of the program that are important to a high level of safety and environmentally sound performance. The committee noted earlier that improvements in nonagent safety could be made. This is an issue of emphasis rather than staffing. The Army is providing the necessary staffing at present to manage the TOCDF adequately during systemization and into the start of agent operations. However, the committee is concerned that the quality of leadership may be affected because several new and "acting" positions have not been permanently filled.

Contractor teams also exhibit good professional capabilities and interact well in support of the office of the PMCD. However, in early May, the EG&G general manager at the TOCDF unexpectedly announced his retirement. The Army and EG&G management are committed to ensuring that a change in this top management position does not compromise safety and environmental standards, or the safety culture established under the leadership of the former general manager.

To ensure continued progress toward the start of agent operations at the TOCDF, a new general manager was appointed at the end of June. He worked closely with the former general manager during the months of July and August, and the latter remains available on a consulting basis to assist in the transition from systemization through the start of agent operations. As the Chemical Stockpile Disposal Program continues to expand in the future, the committee's concerns about staffing will remain relevant.

4

Systemization Environmental Performance

Environmental performance of the Tooele Chemical Agent Disposal Facility (TOCDF) is controlled by compliance with the Resource Conservation and Recovery Act (RCRA) permit. This chapter describes the status of the permit and compliance testing relevant to its requirements.

TOCDF PERMITTING REQUIREMENTS

The TOCDF operates under RCRA permit UT5210090002 issued pursuant to the delegation of authority to the Utah Division of Solid and Hazardous Waste (DSHW) under the Utah Administrative Code, section 315 (R315). Under the requirements of this permit, the incinerator system must demonstrate an ability to treat hazardous waste in a way that protects human health and the environment. Section 3004 of RCRA (1976) requires performance standards establishing the levels of environmental protection that hazardous waste treatment, storage, and disposal facilities must achieve and mandates the criteria against which applications for permits must be measured.

The TOCDF has five incinerator systems that have been scheduled for surrogate testing. These incinerators include the deactivation furnace system, the metal parts furnace, two liquid incinerators (liquid incinerator #1 and liquid incinerator #2), and the dunnage furnace.

The purpose of the surrogate trial burns is to demonstrate the ability of each incinerator system to destroy selected compounds to a destruction removal efficiency (DRE) of 99.9999 percent (6-nines). The compounds are selected to meet criteria specified by the Utah DSHW. The surrogate trial burn for liquid incinerator #1 was conducted between June 30 and July 7, 1995. The deactivation furnace system surrogate trial burn took place in September 1995. Data from these tests must be analyzed and the 6-nines DRE demonstrated prior to initiation of agent incineration in these two incinerator systems. Surrogate trial burns for the remaining three incinerator systems are scheduled to coincide with the use of these systems in agent-destruction operations.

SURROGATE TRIAL BURNS

Liquid Incinerator #1

The surrogate compounds were selected to meet the criteria specified by the DSHW (Downs, 1994). For the surrogate trial burn of liquid incinerator #1, the principal organic hazardous constituents (POHCs) that meet the criteria specified are 1,2,4-trichlorobenzene and tetrachloroethylene. After the successful demonstration of the RCRA permit-required DRE of 99.9999 percent for the defined POHC mixture, the TOCDF will be allowed to incinerate other permitted liquid wastes in liquid incinerator #1.

The Army had concerns about its ability to demonstrate simultaneously 99.9999 percent DRE for both compounds in a mixed feed for two reasons. First, little is known about the potential intermediate combustion products of either compound and their potential interferences. Second, the feed quantities are limited by the thermal capacity of the system, i.e., less of each compound can be fed into the system in a mixed feed than if each compound were fed in separately.

Nevertheless, the Army believed that the liquid incinerator #1 system had the capability of demonstrating the DRE requirements for the POHC mixture. Thus, the surrogate trial burn was conducted with a two-POHC mixture being fed into the incinerator system. The requirements of the surrogate trial burn are to meet or exceed the RCRA limits in 40 CFR 264.343 and to demonstrate DRE for single compounds (not DRE for a two-POHC mixture); if the mixed-feed surrogate trial burn had failed to demonstrate a DRE of 99.9999 percent for either 1,2,4-trichlorobenzene or tetrachloroethylene, a separate trial burn would have been initiated for each POHC that failed.

Although only three successful demonstrations of destruction removal efficiency are required, the TOCDF conducted four POHC burns so that if one of the sample sets was lost or in any way comprised, or if the system operating conditions (including data collection) were not representative of steady-state operations, three valid runs would still be available for analysis.

The liquid incinerator #1 system is designed to meet RCRA performance requirements (40 CFR Part 264). During the surrogate trial burn program, stack emission tests were conducted for oxygen, particulates, carbon monoxide, hydrochloric acid, chlorine, and POHCs. A heat content analysis was conducted on a composite POHC-feed sample once during each POHC burn. A scan of products of incomplete combustion (PIC) of the 40 CFR Part 266 Appendix VIII compounds is being made in place of a scan of POHCs for the supplemental fuel-only run. Tentatively identified compounds will be reported for the 20 largest peaks in the PIC scan.

Four replicate trial burns were conducted. At the conclusion of the fourth run (before any of the results were analyzed) the test director was required to select three runs to be analyzed for the record. The samples from runs 1, 2, and 4 were chosen for analysis, although it is believed that all four runs were equally valid based on the operational data collected during the test. Samples from the third test run were not analyzed.

Table 4-1 summarizes the particulate and hydrogen chloride emissions and DREs for the liquid incinerator #1 surrogate trial burns. The results indicated that the 99.9999 percent DRE was demonstrated and that particulate and HCl emission requirements were met during the test. The complete trial burn test report was submitted to the state of Utah, Department of Environmental Quality, for review and approval on August 25, 1995.

The agent trial burn for liquid incinerator #1 will be done with chemical agents and must demonstrate DREs greater than 99.99 percent (DSHW requirement). Nevertheless, the Army is committed to meeting 6-nines (99.9999 percent) agent DRE for these incinerators.

Deactivation Furnace System Surrogate Trial Burn

The surrogate compounds for the deactivation furnace system surrogate trial burn were selected to meet criteria by the Utah DSHW (Downs, 1994). The POHCs that meet the criteria specified are monochlorobenzene (chlorobenzene) and hexachloroethane.

Although only three successful demonstrations of DRE were required, the TOCDF conducted four POHC burns in the event that one of the sample sets might be lost or in any way compromised, or that the system operating conditions (including data collection) would not be representative of steady-state operations, in which case the affected burn would not be considered a valid, reportable burn. All analytical data from all valid runs were reported. No run was or could be invalidated solely because the target DRE was not achieved. The deactivation furnace system is designed to meet RCRA regulation performance requirements (40 CFR Part 264). During the surrogate trial burn program, stack emissions were tested for oxygen, particulates, carbon monoxide, hydrochloric acid, chlorine, and principal organic hazardous constituents. A heat content analysis

TABLE 4-1 Summary of Results from the TOCDF Liquid Incinerator #1 Surrogate Trial Burn

Parameter	Required[a]	Baseline[b]	Test Run		
			1	2	4
Particulates (gr/dscf)	max. 0.08	0.0026	0.000415	0.000987	0.000531
HCl (lb/hr)	max. 8.76	ND	0.00560	0.00156	0.00691
Tetrachloroethylene (DRE %)	min. 99.9999	N/A	99.9999 980	99.9999 956	99.9999 955
1,2,4-Trichlorobenzene (DRE %)	min. 99.9999	N/A	99.9999 984	99.9999 999	99.9999 940

[a]State of Utah.
[b]JACADS.

was conducted on a composite POHC-feed sample once during each POHC burn. A products of incomplete combustion (PIC) scan of the 40 CFR Part 266 Appendix VIII compounds was made in place of POHCs for the supplemental fuel-only run. Tentatively identified compounds were to be reported for the 20 largest peaks in the products of incomplete combustion scan.

Following the deactivation furnace system surrogate trial burn, the Toxic Substances Control Act research and development trial burn will require a demonstrated DRE of 99.9999 percent or better of the polychlorinated biphenyls in the rocket shipping and firing tubes. The Toxic Substances Control Act trial burn was scheduled for late November or early December 1995.

5

Community Interaction and Planning

The Stockpile Committee report, *Recommendations for the Disposal of Chemical Agents and Munitions* (NRC, 1994c), addressed the committee's concern for citizen involvement in and perceptions of the Army's Chemical Stockpile Disposal Program in the following recommendation:

> The Army should develop a program of increased scope aimed at improving communications with the public at the storage sites. In addition, the Army should proactively seek out greater community involvement in decisions regarding the technology selection process, oversight of operations, and plans for decommissioning facilities. Finally, the Army should work closely with the Chemical Demilitarization Citizens Advisory Commissions, which have been (or will be) established in affected states. There must be a firmer and more visible commitment to engaging the public and addressing its concerns in the program. (REC-6)

Even before this recommendation in 1994, the Army had briefed the Stockpile Committee on a regular basis concerning efforts to establish and implement a public affairs and community relations program for the Chemical Stockpile Disposal Program (CSDP). In the fall of 1994, the Army expanded efforts considerably by establishing the current Public Outreach Program (Busbee, 1994). In addition, the Stockpile Committee adopted a community involvement plan to consider more fully the views of stakeholders and citizens in deliberations, as well as to gather information that could be used as a basis for future recommendations for the Army's public outreach efforts.

With respect to the Tooele Chemical Agent Disposal Facility (TOCDF) and the neighboring communities, the Stockpile Committee has met twice with representatives of the Utah Citizens Advisory Commission (CAC) (March 1994 and 1995), held a public meeting in Tooele (March 29, 1995), monitored various newspaper reports from the area, and met with various representatives of stakeholder and regulatory groups in Utah. In addition, the Stockpile Committee has reviewed a great deal of information provided by the Army concerning public outreach, including the Battelle studies of community viewpoints prepared for Science Applications International Corporation (SAIC) and the Army (Bradbury et al., 1994).

Based on these activities and an analysis of the information received and collected, the committee offers the following observations concerning the Army's efforts to involve the local community and to work closely with the Utah Citizens Advisory Commission, the Chemical Stockpile Emergency Preparedness Program (CSEPP) in Utah, and planned future community communications and citizen involvement programs in the state.

UTAH COMMUNITY INVOLVEMENT

Community outreach can be encouraged through a variety of mechanisms. The Army's current efforts and plans include: grassroots information campaigns, civic organization meetings and public forums, community relations and special events programs such as tours of facilities, the development and distribution of brochures and videotapes, legislative information support programs, speakers' bureaus, coalition building with regulators, efforts to coordinate interagency information campaigns, and crisis information action plans (Busbee, 1994).

The limited resources of the Stockpile Committee have not permitted direct oversight of all of the Army's community involvement efforts. However, the committee did meet with representatives of the citizens advisory commissions from several states in March 1994, including the chair of the Utah Citizens Advisory Commission. In addition, members of the Stockpile Committee met in March 1995 with the Utah Citizens Advisory Commission, along with representatives of the Utah regulatory community and other state and local agencies with programmatic responsibilities resulting from the disposal program. The other agencies included several divisions of the

Utah Department of Environmental Quality, the Utah Division of Comprehensive Emergency Management (CEM), and the Tooele County Department of Emergency Management. Further, the committee held a sparsely attended public forum to hear the viewpoints of community residents.

Based on these discussions and the review of materials provided by the Army, several concerns need to be raised that have potential adverse effects on either the scheduled disposal in Utah or the larger continental program.

Utah Citizens Advisory Commission and Risk Assessment: Problems of Communication

The Congress in October 1992 (Public Law 102-484, Section 172, see appendix A) mandated that the Army establish citizens advisory commissions in each state where low-volume chemical weapons sites existed and, at the request of the governor, in any state where there is a chemical weapons storage site. The Army was further mandated to meet with these commissions to hear citizen and state concerns regarding the disposal program. The citizens advisory commissions were to be appointed by the governor of each state, with seven of the nine members of each commission representing the area near the site and two members representing the state government and having "direct responsibilities related to the chemical demilitarization program" (Public Law 102-484, Section 172(c)).

The Army was informed of the formation of the Utah Citizens Advisory Commission by Governor Leavitt on August 16, 1993 (Leavitt, 1993). The mission of the Utah commission, as specified in its objectives, includes reviewing and advising the governor on issues pertaining to the health and safety of citizens within the state regarding incineration procedures and standards; on issues pertaining to permitting and compliance of the facility with regard to federal and state regulations; with regard to emergency preparedness; on alternative technologies; and on issues relating to the transportation of chemical agents. The first chair of the Utah commission was also the science advisor to the governor.

The Stockpile Committee's March 1995 meeting with the Utah commission revealed that there were communications difficulties between the Army and the commission relating to the site-specific risk assessments being conducted at the TOCDF. The Utah Citizens Advisory Commission indicated that they had not been consulted and that their input had not been solicited by the Army or its contractors for the ongoing risk assessment. The Stockpile Committee's view of the importance of local involvement to the site-specific risk assessments is a matter of record. In a letter (to then Assistant Secretary of the Army Susan Livingstone) on risk assessments for facilities in the continental U.S., the committee stated:

> Local representatives of neighboring communities must be involved early. Their concerns about the CSDP may be substantial, and will warrant consideration throughout the analysis process (NRC, 1993a).

In addition, the committee has pursued the matter of community involvement in site-specific risk assessments with representatives from the Army at almost every meeting since the letter was written. The Utah Citizens Advisory Commission's comment on the lack of Army solicitation of input into the site-specific risk assessment was of considerable concern to the committee because of the importance of this study, not only to the Tooele Army Depot, but also to the other continental sites where site-specific risk assessments are already under way or are scheduled to begin shortly.

The Army has documented numerous occasions when the Army or its contractors explained the purpose of the site-specific risk assessment and solicited input from the Citizens Advisory Commission. For example, on July 27, 1994, Dr. Chris Amos from SAIC provided the commission with an overview of the site-specific risk assessment schedule and solicited input from the commission about perceived risks or scenarios that were perceived to be particularly important (Amos, 1994). The Army also solicited input from the commission chair about the makeup of the risk assessment Expert Panel, and ensured that one member of the panel was from Utah (St. Pierre, 1994). Indeed, in January 1994, the Program Manager for Chemical Demilitarization wrote the chair of the Utah Citizens Advisory Commission advising her of the scheduled risk assessment and suggesting ways the commission might wish to become involved in the process (Baronian, 1994).

Quite clearly, the perceptions of the Utah Citizens Advisory Commission concerning the lack of opportunity for their input into the risk assessment (provided to the committee at the March 1995 meeting) are not congruent with the actions the Army is able to document of solicitation of Citizens Advisory Commission involvement in the risk assessments. The ultimate responsibility for ensuring community involvement in the site-specific risk assessments must rest with the Army. The Army and its contractors met again with the Citizens Advisory Commission and members of the public to discuss the quantitative risk assessment and some of the preliminary results on April 27, 1995. In addition, a public workshop was held in May 1995 to discuss the preliminary results of the risk assessments.

The committee finds the Army's efforts in Utah to obtain community input into the risk assessments were substantial, but not very productive. The committee believes that citizen involvement both prior to and during the work of the risk assessment is essential to improve communications and the ultimate public acceptance of the risk studies.

Personal Protective Equipment

At the Stockpile Committee's public meeting in Tooele, Utah, on March 29, 1995, considerable concern was expressed both by members of the Utah Citizens Advisory Commission and representatives of the Utah CEM about the Chemical Stockpile Emergency Preparedness Program. The single major concern was the lack of Occupational Safety and Health Administration (OSHA) approval of personal protective equipment (masks and protective clothing) for state and local first-responders. The former chairperson of the Citizens Advisory Commission, who was science advisor to the governor of Utah, suggested that the lack of approved equipment constituted a "potential showstopper" for plans to begin incineration because public perceptions would prevent agent burning until the equipment was provided.

The Utah CEM expressed the belief that chemical incineration would not be permitted until personal protective equipment (PPE) is provided to first-responders and they have adequate time to train in the gear. A Utah CEM representative estimated that training would take about six months (Cobb, 1995). Representatives of the Utah Division of Solid and Hazardous Waste indicated that the personal protective equipment is not, strictly speaking, part of the permit process, but they believed the issue needed to be resolved before the start of agent operations (Downs, 1995). The committee believes this issue requires quick resolution because of the continuing larger risk associated with stockpile storage and the need to begin agent operations.

During the committee's March visit, the issue of personal protective equipment for first-responders was clearly critical in the minds of members of the Citizens Advisory Commission, the Utah CEM, and possibly the public. The Army cooperated with local and state officials to try to speed OSHA's approval of alternative personal protective equipment (battle dress overgarment), which Army personnel have used elsewhere. The issue had been on the table for approximately seven years, according to one member of the Citizens Advisory Commission. The director of the Tooele County Department of Emergency Management told the Stockpile Committee that adequate training for the use of personal protective equipment need not take six months, but that failure to complete training could threaten start of agent operations in early 1996 (Sagers, 1995a). As of this writing, OSHA approval for another type of personal protective equipment has been received.

The director of the Tooele County Department of Emergency Management, however, made it clear that the department's reliance on volunteer response personnel would nonetheless not provide for adequate public safety and that county volunteers still needed adequate training in the newly approved personal protective equipment (Lee, 1995). The county department has insisted on federal funding of a "core team" of well-trained responders who would be available to respond to an incident at any time. The importance of this issue can be readily seen in the testimony by the director of the county department to the Procurement Subcommittee of the U.S. House Committee on National Security on July 13, 1995 (Sagers, 1995b):

> Until our public safety needs are met, we cannot in good conscience allow hot operational burns at the Tooele Chemical Agent Demilitarization Facility. Tooele County's position is to let the facility sit out in the desert gathering dust until these matters are resolved.

The Stockpile Committee considers this continuing difficulty to be the result of inadequate integrated planning by the Army of off-site and on-site emergency preparedness and response, and inadequate information about local needs and concerns. The importance of personal protective equipment, the adequate training of personnel, and the response capacity of the community are not new issues. Nevertheless, as of this writing, they threaten start-up of the TOCDF.

COMMUNITY EMERGENCY PLANNING

The Stockpile Committee has been aware of the Chemical Stockpile Emergency Preparedness Program for some time, as well as some of the difficulties the program has encountered (U.S. General Accounting Office, 1993a, 1993b, 1994, 1995). Of particular concern to the committee has been whether communities participating in the program are prepared to respond to off-site emergencies (U.S. General Accounting Office, 1993a). The committee has been briefed by a representative of the new joint office of CSEPP (now under the Program Manager for Chemical Demilitarization) concerning new organizational arrangements between the Federal Emergency Management Agency (FEMA) and the Army (Shandle, 1995). In addition, the Utah CEM also briefed the committee on the CSEPP program in Utah, and Tooele County emergency management personnel conducted a tour of their emergency operations center for the committee on the occasion of the committee's March 1995 public meeting. Finally, the committee has been provided with the Utah and Tooele County Emergency Operations Plans (Tooele County, 1994a) and related CSEPP appendices.

The committee believes it is essential that a coordinated and effective chemical event emergency management capacity be in place as soon as possible because the existence of the stockpile poses a continuing risk. As discussed above, the March 29, 1995, meeting in Tooele revealed to the committee the continuing problem with the personal protective equipment for first-responders. In the unlikely event of a release of agent and an off-site emergency, it is essential that local and state responders be adequately equipped and trained and that they be well prepared through emergency exercises for any eventuality. In addition, potential evacuations, coordinated response, decontamination sites, restoration and reentry activities, and public warning/notification measures all require excellent communication systems.

The committee is not prepared to undertake an exhaustive review of either the adequacy of national CSEPP emergency planning or the various subjects addressed in the General Accounting Office reports. Nor is the committee prepared to address all of the facets of various state and local efforts at emergency preparedness or the coordination in planning efforts between the three Utah counties in the emergency planning zone (EPZ), the state Division of Comprehensive Emergency Management, and the Army emergency response personnel at the TOCDF. Instead, the committee has chosen to state some findings and concerns, based only upon meetings in Utah, review of the emergency preparedness plans from the state and Tooele County, briefings to the committee by Army representatives, and interviews with Tooele County emergency management personnel. These findings are then used as the basis for additional recommendations.

Training

Several issues related to training have been brought to the committee's attention. First, the Utah CEM has pointed out the lack of attention at the national CSEPP level to "critical first-responder operations training" (e.g., personal protective equipment) (Cobb, 1995). The Utah CEM has indicated that, because of a lack of national planning standards, guidance with regard to reentry, emergency medical services, and recovery phase operations has been incomplete and has led to less than effective training and exercises. The issue of reentry is particularly disturbing because Army officials, at the meeting of March 29, 1995, in Tooele, stated they believed reentry would be handled by Army personnel, but Utah CEM officials felt this was a state function. In addition, at the March 29 meeting, a representative from Utah County, one of the three counties in the emergency planning zone, indicated that the county had not participated in the last emergency response exercise because of frustration over the issue of personal protective equipment. In follow-up discussions with representatives of the Tooele County Department of Emergency Management, it was learned that both Utah and Salt Lake counties have played minimal roles in exercises for the same reason (Rutishauser, 1995a).

Exercising emergency response plans is critical to preparedness and actual response. County participation is at their discretion. But, the lack of participation by some counties raises concerns about their level of preparedness for an emergency. In addition, the Utah CEM has raised the issue of whether there are adequate national guidelines for training in several other areas. The Army and FEMA have divided responsibilities for providing adequate training, exercises, and preparedness guidelines. FEMA, according to a Memorandum for the Record signed in February 1994, supports the Army by "working with state and local governments in developing off-post emergency preparedness plans, upgrading response capabilities, and conducting necessary training" (FEMA and the Department of the Army, 1994). These issues were unresolved as of this writing (Lee, 1995).

Emergency Planning

The Army and FEMA have provided detailed CSEPP planning guidelines to state and local communities in many important areas of emergency planning (FEMA and the Department of the Army, 1994). In the event of emergency, off-site efforts are the responsibility of the appropriate state and local governments. The guidance provided by the national agencies suggests that CSEPP emergency plans should be appended to the existing all-hazards emergency plan of state and local governments as functional appendices specifically geared to chemical agent response (FEMA and the Department of the Army, 1994). The committee has had the opportunity to review the Tooele County Emergency Operations Plan and the functional appendices directed to chemical agent response (Tooele County, 1994a). In addition, the committee has been briefed and has toured the Tooele County emergency operations center.

The review of the Tooele County functional appendices dealing with chemical agent incidents and intended for emergencies related to the Tooele South Area raises several concerns. First, many of the draft appendices were only completed in the fall of 1994 and were still in draft form as of April 18, 1995 (Rutishauser, 1995a). Second, probably because they are still drafts or preliminary in nature, some parts of the appendices are not well integrated. The appendices available for committee review included: Communications; Fire/Rescue; Health and Emergency Medical Services; the Community Command Post Concept; the Emergency Operations Plan; and Disaster Reception Center Standard Operating Procedures. In addition, the CSEPP program manager has indicated that the Health and Emergency Medical Services Appendix is in "very draft form." The Reentry and Restoration Appendix was still in progress as of February 11, 1995 (Rutishauser, 1995b).

Consistent with Army and FEMA guidelines, the committee believes it is essential that these plans be fully prepared and exercised in order to provide complete and comprehensive emergency response capability to a chemical event (FEMA and the Department of the Army, 1994). The finding of the committee is that the local CSEPP emergency planning efforts are not complete, as evidenced by the draft version of the appendices.

Emergency Communications

As FEMA and the Army have pointed out, "the emergency communications system must have both a high reliability factor and redundancy." A communications network consisting of redundant telephone and radio systems should be designed and installed to link the Army installation Emergency Operations Center (EOC) and notification point with the EOCs and notification points of all immediate response zone (IRZ) counties and states (FEMA and Department of the Army, 1994).

Effective communications systems are essential for notification and warning, incident command, emergency operations center functions, linkage to state and Army EOCs, linkage to sheltering centers and decontamination sites, communications with local elected officials, and linkage to all immediate response zone first-responders. The importance of effective communications for disaster response and effective public warning and evacuation is well documented in the disaster and emergency response literature (FEMA and Department of the Army, 1994).

The Tooele County Communications Plan is a comprehensive document despite its preliminary nature. The plan allows for redundancy and provides sites for coverage of the CSEPP counties to include microwave, VHF, 400-MHz, 800-MHz, and 900-MHz channels (Tooele County, 1994b).

Indoor alert and notification system radios (990 Tone Alert Radios) have only recently been funded and were expected to be in place in the community in the fall of 1995. In addition, Phase II of the 800-MHz system, which was scheduled for completion in 1994, has been delayed again and was not funded by FEMA in the FY 1995–1996 budget received by Tooele County (Lee, 1995). This system was to have enabled, among other functions, direct communications with the other two counties in the emergency planning zone and will require some additional training of personnel. In addition, some repeaters for Salt Lake and Utah counties have been delayed because of funding, and as a result, communications are problematic in parts of southern Tooele County, including Rush Valley (McCall, 1995).

In short, the committee has found that the communication plan for Tooele County and the planned implementation of the communications system linking important operations centers in the emergency planning zone are not yet complete. The Army has not yet authorized funding for implementation of the communications system (Lee, 1995).

Emergency Medical Care

The Tooele Valley Regional Medical Center is the primary local medical facility and is working with the Army on issues of emergency medical service. The medical center has asked the Army to provide funding to keep the hospital emergency room open on a 24-hour basis and to train hospital personnel in procedures applicable to agent exposure. In congressional testimony, the director of the Tooele County Department of Emergency Management stated her concerns over delays in funding as well as the small fraction of the Army's CSEPP budget that has been allocated in Utah (Sagers, 1995b). She stated that Tooele County is not presently prepared to respond appropriately to an emergency involving agent release.

ARMY CITIZENS INVOLVEMENT PROGRAM IN UTAH

The Stockpile Committee has stressed the importance of citizen and community involvement not only in the risk assessment process but also in decisions relating to the technology selection process, oversight of operations, and plans for decommissioning (NRC, 1994c). Involvement of citizens and the community in the selection of technology at the TOCDF was not possible because incineration had already been selected for the site by the time the committee made its recommendation. Based on briefings by Army personnel, the Stockpile Committee meetings in Utah with citizens, and reports prepared for various governmental entities, the committee can now comment on several aspects of the Army community involvement program in Utah.

Consistent with the memorandum and plan sent from Brigadier General Walter Busbee to the principal Deputy Assistant Secretary of the Army on September 30, 1994 (Busbee, 1994), a public outreach program has been developed and should be implemented in Utah. The overall Army outreach effort consists of at least the following: community tours of the TOCDF; media tours of the TOCDF; CSEPP tours; state regulator tours; and tours for legislators and other governmental officials. The following educational materials are also planned: production of a systemization video; fact sheets for distribution to community and civic groups; speaking package materials; and educational materials for schools. Outreach activities include: planned risk assessment workshops; planned outreach to the schools; special events, including tours and community events; planned speakers' bureaus; expanded and updated mailing lists; and involvement of CSEPP representatives in community outreach activities.

The committee has also been briefed on the hiring of additional staff for the program in Tooele, including a storefront office (Fournier, 1995). The Army also contracted with Battelle Pacific Northwest Laboratories through Science Applications International Corporation to conduct a series of interviews and focus groups in Tooele and at the other depot sites to better understand how people interpret the demilitarization program, "why they respond in certain ways and what the Army could do to design effective ways to interact with the public" (Bradbury et al., 1994). Finally, the Tooele County Department of Emergency Management contracted with Insight Research in Salt Lake City to conduct a survey of Utah and Tooele county residents on issues related to CSEPP and emergency management (Insight Research, 1994). This report indicates that about 20 percent of the residents felt they were familiar with the program, 20 percent had heard the name only, and 60 percent had never heard of it.

The Battelle study in Tooele indicates that residents want the chemicals destroyed and are concerned about future use of the incinerator at the TOCDF, the transport of chemical weapons from other sites, and long-term health and environmental effects from environmental contamination at Tooele, Dugway Proving Ground, and other commercial incineration facilities nearby. In addition, there is substantial concern over cumulative environmental and health effects from incineration at the TOCDF and the other facilities (Bradbury et al., 1994).

The Stockpile Committee efforts to solicit public input in Tooele were disappointing (i.e., citizen turnout was low). Although numerous invitations were mailed (appendix D), only four citizens from the area appeared on March 29, 1995, to comment on the TOCDF and other elements of the program. One citizen commented that local residents trust the people working at Tooele Army Depot. Another individual, a member of a group associated with the Chemical Weapons Working Group, expressed concerns about the potential change in requirements in the agent trial burn from 6-nines (99.9999 percent) to 4-nines (99.99 percent). Other issues raised by this individual related to the long-term effects of products of incomplete combustion, the possibilities for reconfiguring the stockpile by draining the agent first and then using neutralization (chemical hydrolysis), and what was perceived as a National Research Council (NRC) bias toward incineration. Another citizen, a member of the Sierra Club, raised issues related to the 4-nines for trial burns, neutralization alternatives to incineration, CSEPP's evacuation planning, liability issues related to hazardous materials teams that might aid in an emergency, and, finally, the lack of decontamination equipment in the county hospital facility. There remained some ambiguity about the expertise of the county in terms of both hazardous materials teams and equipment and decontamination units other than mobile facilities.

After reviewing these materials and after public discussions, the committee finds that the Army has begun to implement a large and comprehensive public information program in Tooele County. Considering the list of activities either planned or under way in this program, it is clear that insufficient attention may have been given to the importance of soliciting citizen input into programmatic decisions as outlined in Committee Recommendation 6 (REC-6) in the *Recommendations* report (NRC, 1994c). For example, the committee found no evidence that the Army has considered the role of the public in decisions related to decommissioning the facility or monitoring disposal operations. It is understandable that the Army's program is oriented toward producing and disseminating information to the public, but it is essential that the outreach also proactively develop mechanisms and an approach for "engaging the public and addressing its concerns in the program" (NRC, 1994c) (Hance et al., 1988). Public concern about emergency management in the program is likely to increase as the TOCDF starts agent operations. Efforts should be made by the Army to include the public in CSEPP issues and to integrate public outreach more fully with elements of the CSEPP program.

6

Overview of Site-Specific Risk Assessment

The National Research Council's (NRC) Stockpile Committee was established to provide the Army with independent scientific and technical advice on the Chemical Stockpile Disposal Program (CSDP). In carrying out its activities, the committee has maintained a continuing interest in the scope and methods employed by the Army to manage the risks associated with storage and disposal of the stockpile.

In the past, the committee reviewed activities at Johnston Island (NRC, 1984) in the Pacific Ocean, about 700 miles southwest of Hawaii, where the Army completed construction in 1990 of the first full-scale facility for the destruction of the stockpile using the baseline incineration system. Reviews of the Johnston Atoll Chemical Agent Disposal System (JACADS) Operational Verification Testing (including destruction of M55 rockets with GB and VX, ton containers of mustard, and 105mm artillery projectiles with mustard) (MITRE, 1993c); the JACADS-specific risk assessment (U.S. Army, 1987) and the risk assessment performed in support of the Army's Final Programmatic Environmental Impact Statement (FPEIS) (U.S. Army, 1988); on-site observations of JACADS operations; and visits to the Tooele Chemical Agent Disposal Facility (TOCDF) construction site led the committee to recommend additional risk assessment work.

The committee noted that the risk assessment to support the FPEIS was directed toward assessing the relative risks among various disposal (or continued storage) options (NRC, 1993a). The risk analysis as presented in the FPEIS was not directed at managing risk at any one site. The continental sites at which lethal chemical agents and munitions will be destroyed all differ substantially from Johnston Island, as well as from one another, with regard to terrain, weather, the density of nearby population, transportation network, the size and variety of stored agents and munitions, other aspects, and, possibly, destruction technology. The committee observed that the FPEIS needs to be supported by more site-specific risk analysis to provide an adequate basis for total risk management of the CSDP.

Experience in the nuclear power industry has demonstrated that risk assessments provide the best basis for quantitative risk management (U.S.Nuclear Regulatory Commission, 1994) (Taylor, 1994). In addition, they have been shown to provide a rational basis for the discussion of community concerns. Knowing not only what the risk is but also what is driving it at the basic event level permits specific actions of risk control and a quantification of the effectiveness of those actions.

The committee stated that the long-range benefits to the Army of site-specific risk assessments goes beyond simply quantifying the risk. The site-specific risk assessment can be an important management tool in the decision-ranking process in relation to design, operations, personnel training, maintenance, and plant modifications. Site-specific risk assessment is necessary at every location where lethal chemical agents and munitions will be destroyed, regardless of the method of destruction.

This chapter presents the committee's specific previous recommendations with respect to risk assessment and risk management and the committee's interpretation of the Army's response to those recommendations. It describes the Army's plans for risk assessment and risk management at the TOCDF and Tooele Army Depot chemical stockpile and their current status. Finally, it provides the committee's findings with respect to the chemical agent disposal facility risk management program.

NRC RECOMMENDATIONS ON RISK MANAGEMENT

The committee has presented recommendations on risk management of the CSDP in three separate documents. Although there is some duplication among the recommendations, all of them are discussed in the following pages.

Report on Operational Verification Testing

The final recommendation in the Stockpile Committee's *Evaluation of the Johnston Atoll Chemical Agent Disposal System Operational Verification Testing: Part II* (NRC, 1994a) dealt with risk assessment:

> Complete the risk assessment for the Tooele Chemical Agent Disposal Facility during the systemization period. (OVT2-6)

Because experience from Operational Verification Testing (OVT) at JACADS resulted in a number of changes to the design and planned operations at TOCDF, the committee wanted to ensure that the effects of those changes on risk would be examined. The Army has adopted this recommendation in stepwise fashion. Because of time constraints, the entire risk assessment will not be completed before the start of agent operations at the TOCDF. However, risk assessment pertinent to each campaign (the disposal of a particular agent/munitions type) will be completed and lessons learned will be applied before the start of each campaign. The first analysis for campaigns 1 and 2 (GB and VX M55 rockets with coprocessing of bulk items) was completed in April 1995. Modifications to the TOCDF and its operations are in progress and will be completed prior to the start of agent operations (St. Pierre, 1995a). The draft report describing the final analysis for the first two campaigns was completed on June 26, 1995. The complete TOCDF risk assessment is expected to be published in the first quarter of 1996. The committee believes that the Army's approach to the TOCDF risk assessment meets the intent of this recommendation.

NRC Letter Report on the Chemical Stockpile Disposal Program Risk Management Process

Recommendations for Facilities in the Continental United States

In January 1993, the committee sent a letter report to the Assistant Secretary of the Army recommending specific actions to enhance the CSDP risk management process (NRC, 1993a). That letter report made three specific recommendations for site-specific risk assessments for facilities in the continental United States. The recommendations are discussed below. The first was:

> A site-specific, full-scope, scenario-based risk assessment should be performed for each continental U.S. facility, starting with the Tooele facility. (RISK-1)

The Army's site-specific, full-scope, scenario-based risk assessment of the TOCDF is described in the following sections. For technical and administrative reasons, there are multiple risk assessment projects and reports as described below (e.g., "an accident quantitative risk assessment," "a health effects risk assessment," and "a reconfiguration risk assessment"). The committee understands the logic behind the various assessments but believes that the multiplicity of assessments can cause misunderstanding among reviewers, government agencies, and the public. The Army should adopt a standard language that recognizes the ensemble of risk-related projects as "the risk assessment," and individual studies should always be referred to as components of the wider "risk assessment."

The second recommendation was:

> Each site-specific risk assessment should include the case of continued storage without disposal as one scenario. (RISK-2)

The current Tooele risk assessment is limited to analysis of the activities associated with the first and second campaigns. Risks associated with continued storage will be addressed later (St. Pierre, 1995b). The accident quantitative risk assessment (QRA) team obtained additional expert assistance to address specific aspects of the risk of continued storage and associated agent/munitions handling. An expert seminar was held on June 27–30, 1995, to develop better models and data for evaluating risks from handling munitions. In addition, an external expert review of health effects and consequence modeling was held on June 22, 1995, to provide a better basis for modeling these effects.

The third recommendation was:

> The risk assessments should be quantitative and include the following features:
>
> - bottom-line results on the health effects to on-site personnel in terms of likelihood and consequence, including a site-specific atmospheric dispersion and health effects analysis and an analysis of emergency response capability;
> - a clearly defined set of scenarios that, taken together, provide a comprehensive representation of the risk;

- dependency matrices that display inter- and intra-system dependencies;
- a human action analysis that represents the human role in controlling risk;
- quantification of risk from all causes, including both internal events (plant and plant-people failures) and external events (earthquakes, fires, floods, aircraft crashes, etc.);
- site-specific hardware, software, procedures, training programs, maintenance practices, and operations personnel (including site-specific storage facilities and munitions handling activities);
- risk contributors in such terms as random failures, common cause failures, multiple failures, and human error; and
- an uncertainty analysis to display clearly how much confidence the analysts have in the precision of the quantitative results. (RISK-3)

The accident QRA is progressing in a manner consistent with the goals of this recommendation. Not all the above features are included in the summary report on campaigns 1 and 2, a draft of which was circulated for comment on June 26, 1995, but they are all part of the ongoing accident QRA.

Broad Guidelines for Risk Management

The committee's letter report provided five broad guidelines the Army should use to conduct site-specific analyses. The first guideline was:

Modern, up-to-date methodologies should be employed, such as those found in the risk assessments reported in NUREG-1150. (RISK-4)

The accident QRA meets the first guideline. The second guideline was:

The risk assessments should be conducted by organizations with recognized expertise in the field, but not otherwise involved in the CSDP. In a similar vein, independent peer reviews are an absolute requirement. (RISK-5)

Science Applications International Corporation (SAIC), the organization conducting the analysis, has recognized expertise in the field, as do the individuals from SAIC actually leading the work. In areas where SAIC does not have the requisite expertise, they have obtained the services of subcontractors who meet the requirements. In a few difficult areas where methods have not yet been developed and data are sparse, SAIC is convening expert panels to develop recommendations based on experience and expertise.

SAIC is involved in other projects at the TOCDF and for the Chemical Stockpile Disposal Program. Nevertheless, the Army and other organizations within SAIC appear to be protecting the QRA team from outside influence and to be dealing with them as an independent organization. Discussions with team members show that they have a healthy skepticism for information and opinions expressed by the Army management and by the Army, EG&G, and SAIC personnel involved with the TOCDF. The evolving process for risk management reflects a challenging and competitive, but still cooperative, relationship between these organizations.

Independent review has been established with the QRA Expert Panel, a group of five experts brought together under contract with MITRE and who operate independently of project management. Three of the panel members have extensive QRA experience, primarily in the field of nuclear reactor safety but with additional experience in the analysis of aerospace and process chemical facilities. One panel member is a member of the U.S. Nuclear Regulatory Commission's Advisory Committee on Reactor Safeguards. Another is a combustion expert from Brigham Young University in Salt Lake City, who also provides some degree of local perspective for the panel. Two are professors of engineering at major universities. Two are chemical engineers with process safety experience. All five have extensive professional experience and are consultants for major organizations. One member of the Stockpile Committee member was invited to monitor the Expert Panel meetings, generally two days each at two-month intervals. The Stockpile Committee member provided no guidance (that is the role of members of the Expert Panel), but was able to interrogate individual analysts and to delve deeply into the details of the QRA analysis. The committee is satisfied that the approach taken by the Army meets the intent of the recommendation.

The third broad guideline was:

Local representatives of neighboring communities must be involved early. Their concerns about the CSDP may be substantial, and will warrant consideration throughout the analysis process. (RISK-6)

The Army attempted to involve the community as early as spring of 1994 (Amos, 1994) and at intervals since then. However, as was noted in chapter 5, key elements of the community seem to have missed or misinterpreted the interactions. Additional efforts are required to involve the community successfully. Lessons learned here could be applied at future sites. The fourth broad guideline was:

> Emphasis must be placed on human reliability factors, particularly in light of the human factors issues raised by the Stockpile Committee in reviewing the first phase of Operational Verification Testing at JACADS. (RISK-7)

The QRA team has brought human reliability experts into the project. However, so far, because of the limited quantity of agent that can be processed in the facility, no human error event significant to off-site release has been identified. The committee expects that the impact of human reliability on worker risk and stockpile handling accidents could be important, and the committee expects that the QRA will examine these areas. The committee will carefully monitor this analysis as the work proceeds.

The fifth broad guideline was:

> To avoid overstatement of the results it is important that the confidence levels of the risk parameters be fully displayed. It is this process of quantifying the uncertainty in the risk that will establish the reliability of the conclusions. Experience has indicated that the results of a risk assessment provide valuable information on the importance of different contributors to risk, not only in terms of hardware failures but also in terms of human errors and deficiencies in procedures and software. Thus the risk assessment can lead to process changes that reduce overall risk. (RISK-8)

The Expert Panel had asked the QRA team to expand and clarify their effort in characterizing the uncertainty. The committee expects that the QRA will meet the intent of this recommendation and will examine this area as the work proceeds.

Recommendations Report

In the *Recommendations* report (NRC, 1994c), the Stockpile Committee stated that risk analyses of both storage and disposal operations have shown that cumulative total risk to the public and to the environment is generally dominated by storage rather than disposal operations, at least in terms of the risk of acute agent exposure. The committee suggested that the risk analyses be updated, but recognized that strong evidence shows that new studies are not likely to alter significantly the distribution of risk. Furthermore, although there is no evidence of imminent disaster, storage and disposal risks will increase in time as the stockpile continues to deteriorate. The only way to avoid the continuing and growing risk of both acute agent exposure and long-term health effects is to eliminate the hazardous materials. Recommendations 2 through 5 of the *Recommendations* report deal with risk assessment. The recommended assessments were intended to verify the conclusions for each specific site and to address concerns expressed by the public.

The first of these recommendations was:

> The committee expects the latent risk from storage, handling, and disposal activities to be low. However, new risk analyses should be conducted that explicitly account for latent health risks from storage, handling, and disposal. (REC-2)

The latent risk from agent release during disposal activities (including handling required for disposal) for campaigns 1 and 2 was calculated in the draft summary report circulated June 26, 1995. Latent risk from continued storage and handling, as well as risk from later campaigns, will be included in the accident QRA that is to be published early in 1996. Latent risk from routine releases during normal operations is the subject of the health risk assessment to be directed by the state of Utah. The committee is concerned that this study may not be fully site-specific (e.g., perhaps using meteorology from Salt Lake City rather than Tooele), may not be directly comparable with the accident QRA results, and may not realistically address the probabilities of various possible releases. The Army should link the health risk assessment results with the accident QRA in one summary document describing the entire Tooele risk assessment and risk management program.

The second of these recommendations was:

> Updated analyses of the relative risk of storage, handling, and disposal activities should be completed as soon as possible. (REC-3)

The updated analysis for accidental release of agent at the TOCDF is under way and is expected to be completed in early 1996. Although the analysis of the risk of storage and continued handling has not yet been completed, the campaign 1 and 2 accident QRA suggests that expediting destruction of agent will reduce total risk to the public. Earthquake scenarios are major contributors to risk at the TOCDF (Benjamin, 1994). Seismic events are likely to pose even greater risk to the stockpile than to the disposal facility. Therefore, the longer the stockpile remains, the higher the cumulative likelihood of such an event.

The third recommendation was:

> New risk analyses should be site specific, using the latest available information and methods of analysis. At this time, since there is insufficient knowledge of potential alternative technologies, a first-cut series of analyses should compare the relative risks of continued storage and disposal by the baseline system. Analyses should identify the major contributors of total risk including storage. The analyses will confirm or refute the present belief that maximum safety dictates prompt disposal. (REC-4A)

At this writing, no more than a first-cut analysis of storage risk is available for the TOCDF; a complete analysis is expected in early 1996. For reasons described above (REC-4A), it appears that maximum safety dictates prompt disposal. The Army should revisit this issue as the risk assessment for the TOCDF and chemical stockpile proceeds.

The fourth recommendation was:

> As new, site-specific risk analyses become available, the Army should reconsider the schedule of construction and operation of disposal facilities and, if indicated, reorder the remaining sequence so as to minimize any subsequent cumulative total risk. The Army should also consider reconfiguring each high-risk stockpile to a safer condition prior to disposal if this will significantly decrease cumulative total risk. (REC-4B)

Risk management of the stockpile is planned to include reconfiguration alternatives. Some reconfiguration has been accomplished at another site but not yet at the TOCDF. The Army has recently indicated that the TOCDF site manager has requested inclusion of the stockpile (i.e., continued storage) in the TOCDF Risk Management Plan discussed below. The committee strongly supports this idea.

The fifth recommendation was:

> As research progresses on potential alternative technologies and as their potential for improved safety becomes apparent, site-specific risk analyses should be reexamined, with the potential alternative substituted in the baseline system, to estimate overall system performance. In view of the limited potential for overall safety improvement, however, the disposal program should not be delayed pending completion of such research. (REC-5)

Work on alternative technologies is continuing, as is construction and operation of the baseline incineration system. It is still too soon to perform comparative risk assessment for any alternative technology. A baseline incineration system exists at the TOCDF, and alternative technologies are not under consideration for this site. Once proven alternatives become available, they may be considered if indicated by the integrated programmatic risk assessment for Tooele.

TOCDF RISK MANAGEMENT PLAN

The reason for performing risk assessment is to permit a more rational basis for managing facility design and operations. To take advantage of a thorough risk assessment, it is essential to develop a risk management plan that lays out the process for using the risk assessment within the overall plant management structure. The Army produced the first draft of the TOCDF Risk Management Plan (RMP) in April 1995 (U.S. Army, 1995c). The RMP is written for the TOCDF site manager, but some material was added for other audiences. The TOCDF RMP is intended to be the central guidance document for plant safety, integrating the myriad regulations that define safety at TOCDF. Indeed, one purpose of the RMP is to demonstrate the wide variety of programs in place that combine to enhance safety at the facility. Figure 6-1 (U.S. Army, 1995c) illustrates the logic flow of the RMP, with the accident QRA playing a central role in the decision process for emergency preparedness, management of change, performance evaluation, and incident investigation, and with all these activities feeding back to the QRA. Figure 6-2 (U.S. Army, 1995c), depicts the range of federal (OSHA, EPA, and DoD), Army, USACDRA (PMCD), Tooele Army Depot, Utah, and local regulations that drive the TOCDF safety program. These regulations will not be fully integrated into the body of the RMP. But they are tied together, and the separate documents are cited. Recent indications are that the Army expects to integrate the TOCDF and Tooele Army Depot stockpile programs into the RMP. The draft is currently being revised for

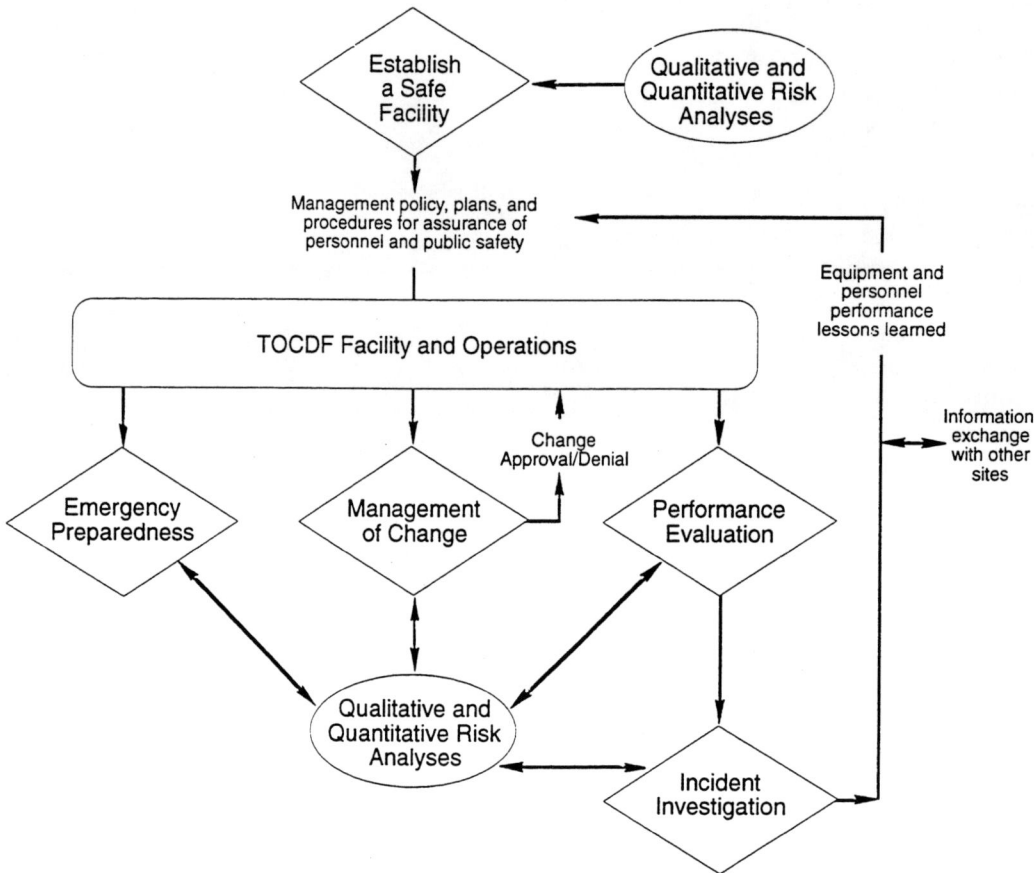

FIGURE 6-1 Overview of the Risk Management Plan. Source: U.S. Army, 1995c.

consistency and completeness and is expected to be in place before the start of agent operations. The committee endorses the RMP approach and the integration of the TOCDF and site risk management.

The committee has also observed the risk management process in action when the Army reviewed early accident QRA results, developed alternative plans for dealing with high-risk scenarios, evaluated them, and adjusted schedules for operations and actual plant hardware to eliminate unnecessary risks. Formalizing an effective process of continuing risk reduction in the RMP is an important goal. The examples below are from a recent TOCDF memorandum (Holmes, 1995). Risk results leading to these changes were calculated in the first draft analysis (U.S. Army, 1994d).

During the development of the QRA, there has been extensive interaction between the QRA team and facility personnel (both on site and in the headquarters of the Program Manager for Chemical Demilitarization). This interaction began with development of process operational diagrams (PODs), which were used to define systematically the process steps. Review of these diagrams by both the QRA team and project personnel increased the operational understanding of both parties.

As the QRA nears completion, routine reviews of insights are held between the QRA team and project personnel. These reviews serve many purposes. If the assumptions in the analysis are not correct or are inconsistent with experience, the assumptions are corrected and the assessment recalculated. If the analysis warrants, discussions are held to identify modifications to the operating strategy or operating procedures to mitigate the risk. The QRA frequently identifies additional processing step modifications that would most effectively mitigate the risk. Assumptions are jointly developed and used in the assessment to ensure proper mitigation of the identified risk. If the analysis results are acceptable, the approach agreed upon is implemented in the field, and the assumptions are published

OVERVIEW OF SITE-SPECIFIC RISK ASSESSMENT

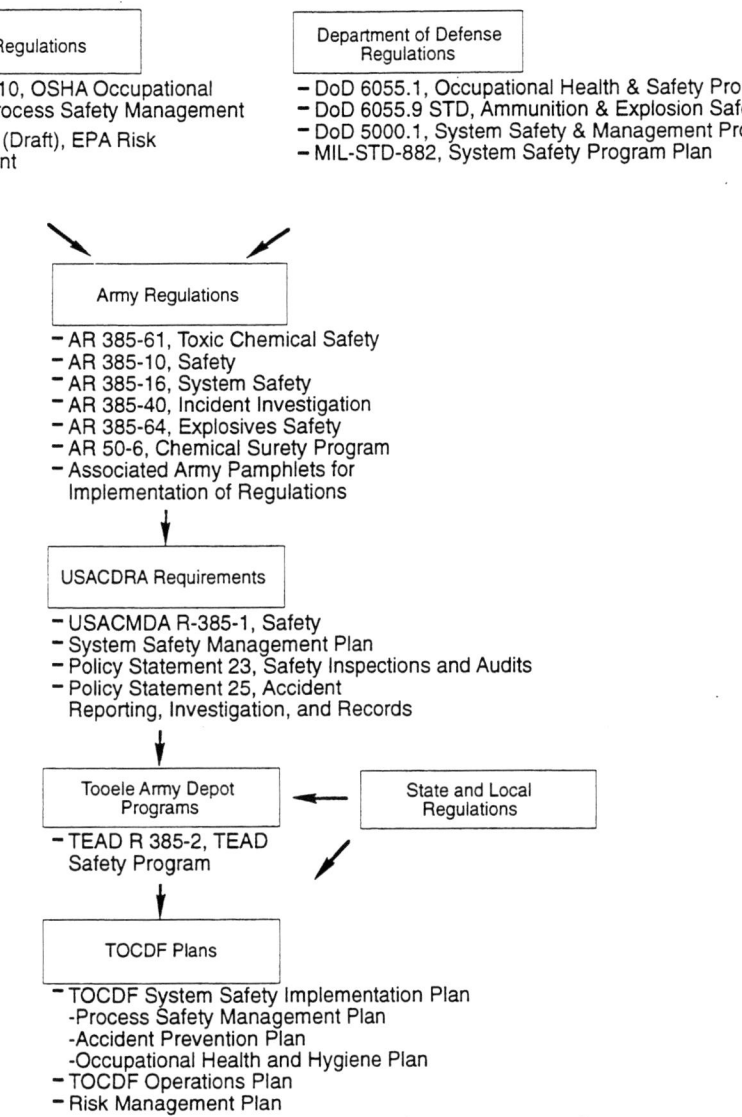

FIGURE 6-2 Hierarchy of regulations that define safety at the TOCDF. Source: U.S. Army, 1995c.

as part of the risk assessment documentation so they can be validated during the site Pre-Operational Survey.

Interaction between project personnel and the QRA team is informal at this point. Nevertheless, the approaches utilized will serve as the basis for developing the analysis requirements documented in the Risk Management Plan.

As discussed earlier in this report, the QRA has already affected several aspects of TOCDF operations. Some of these are described below (see also chapter 2).

Operation of Metal Parts Furnace Feed Airlock

The QRA identified a possible scenario involving the buildup and ignition of vapors in the metal parts furnace feed airlock. The phenomenon was most significant for bulk agent containers but also applied to projectiles. As a direct result of this insight, steps are being taken to limit the possibility that this event could occur, possibly by venting the airlock to the metal parts furnace afterburner. Procedural changes could also significantly limit the probability of this type of scenario. The QRA team is working with the Program Manager for Chemical Demilitarization to assess proposed modifications to minimize the risk. This issue will be resolved before start-up of metal parts furnace agent operations.

Weteye Bomb Aluminum and Agent Interaction

The QRA identified the potential for interaction between molten aluminum and liquid agent in the metal parts furnace during processing of the aluminum weteye bombs, possibly leading to an explosion within the furnace. Calculations in support of the QRA indicated that the presence of liquid agent at the onset of aluminum melting could not be ruled out. The potential for the explosion of molten aluminum and water is known to exist, although there is a good deal of uncertainty about the conditions that produce explosions. No information is available concerning aluminum and agent, but the phenomenon could not be ruled out.

As a result of this finding, the order of the campaigns was changed. Ton containers will be co-processed with GB rockets during the first campaign instead of weteye bombs. This change was made for two main reasons. First, the metal parts furnace GB trial burn requirements would require larger than normal quantities of agent to be left in each weteye bomb to establish the environmental permit feed limits to the furnace. This would further increase risk levels over the preliminary level in the risk assessment. Second, delaying weteye processing will allow time for a review of the SAIC calculations and the development of processing strategies to prevent molten aluminum-agent interaction. Assumptions developed between the QRA team and the Program Manager for Chemical Demilitarization will be the basis for the risk assessment and processing strategy development.

Weteye Bomb Handling and Inventories

The QRA found that weteye bombs increase risks significantly because they contain GB and are relatively thin-walled. The QRA developed a sensitivity study to determine the number of bombs to be loaded on a truck during transport. Additional analysis in the QRA may indicate the number of weteye bombs that should be stored in the container handling building to minimize the risk.

Seismic Anchorage of the Liquid Propane Gas Tank

The facility review of equipment fragilities during an earthquake indicates the liquid propane gas tank anchorage has some seismic vulnerability despite its construction to seismic zone three requirements. Failure of this tank during an earthquake contributes significantly to seismic risk at the facility. The information provided by the QRA team will be used for evaluating the need for additional bracing for this tank.

TOOELE RISK ASSESSMENT

The Army selected SAIC to perform an accident QRA for the TOCDF and the stockpile. That risk assessment follows the general approach of NUREG-1150 as suggested in committee guideline RISK-4 of the letter report on the CSDP risk management process (NRC, 1993a), i.e., an assessment of risk from accidents at the site. But there is more to the risk assessment recommended by the committee than an accident QRA, although previous experience indicates that the principal risk to the general public comes from accidents. At the present time, the Army's risk assessment of the TOCDF has three components:

1. *Accident quantitative risk assessment.* This assessment quantitatively analyzes the probability and consequences of accidental releases of agent at the TOCDF and the chemical stockpile from all conceivable accidents at the Tooele Army Depot. Both acute and latent effects of acute releases are calculated for both workers and the general public. This work is in progress and the analysis has been completed for campaigns 1 and 2. The accident QRA will be described in detail in the following section.

2. *Health risk assessment.* This assessment quantitatively analyzes, generally following Environmental Protection Agency (EPA) guidelines, the possible doses to the general population due to normal operations of the disposal facility. An assessment of the impact on agriculture will also be included (St. Pierre, 1995c). The state of Utah will conduct this analysis. It is anticipated that the assessment will use the approach developed by the Army in the analysis of the Anniston Chemical

Activity (U.S. Army, 1995d). The health risk assessment is not yet available for review and is not discussed further in this report.

3. *Reconfiguration risk assessment.* After the risk from accidental releases of agent during continued storage is calculated, alternative storage configurations will be considered and the risk from each compared. The management structure for this work is not currently in place.

The committee has several concerns about the current constitution of the TOCDF risk assessment. First and foremost, the studies are being performed as separate projects under diverse management with no obvious coordination. The committee believes that a single "TOCDF Risk Assessment Summary Report" can be prepared to present the risk assessment in an integrated fashion as a single, all-encompassing document to the public. From a summary report, people with diverse expectations can easily identify where and how concerns have been addressed. Furthermore, planning for an integrated document will require identifying areas where management's attention is required to permit useful combining of results. State and Army management teams ought to agree on the key issues that must be properly handled in each risk assessment element. For example, each element should be consistent in all parameters at the Tooele site that can affect risk, e.g., design, operation, weather, and demographics.

ACCIDENT QUANTITATIVE RISK ASSESSMENT

The accident QRA of the TOCDF and the Tooele Army Depot stockpile is being performed for the Army by SAIC, a company whose analysts have extensive experience in quantitative, probabilistic risk assessment. The workplan for the QRA is presented in an overview document (Brandyberry, 1994). In some areas, SAIC has engaged subcontractors with special expertise, e.g., analysis of seismic hazards, structural mechanics, latent health effects of agent, and munitions fragility (if dropped). As discussed earlier, the QRA team is essentially independent of outside influence. Risk management activities with respect to scenarios identified in the analysis are performed by the Army in consultation with the QRA team.

The goal of the study is to consider all possible upset conditions during normal and special operations that could lead to release of agent, both within the site boundaries and outside. The analysis is known as a level 3 analysis, because it will calculate acute and latent health effects of agent exposure for workers and the general public. A level 1 analysis focuses on the probability that equipment and human failures can create the potential for a release. A level 2 analysis characterizes the release, and a level 3 analysis calculates the consequences. As discussed earlier, the Army has engaged an independent panel to review and ensure the quality of the QRA.

Methodology

The accident QRA is being performed using a modified form of the standard fault tree/event tree QRA modeling techniques originally developed in the reactor safety study of Rasmussen (Rasmussen, 1975). Details are presented in a methodology manual (U.S. Army, 1994e). Process accident flowchart models, called process operational diagrams (PODs) in the QRA, have been devised as the mechanism for encoding process information. Upsets are identified in the POD as demonstrated in figure 6-3. A POD is described in the Army manual (U.S. Army, 1994e):

> A POD is a step-by-step search for events and upsets...By asking a set of what-if questions after each successive operational step, a thorough assessment of potential upsets can be generated. During this process, existing analyses are referenced to ensure that previously suggested events are covered... [Start by] listing the major steps of the normal operations...Given each normal step, it is necessary to consider all deviations that could occur during that step or if that step did not happen properly...The PODs are used to document the steps in the process and allow efficient review by operational staff.

This format is understandable to engineers and operators familiar with the process who may not be versed in QRA modeling techniques.

To illustrate POD modeling, figure 6-4 traces the major activities in each of the disposal processes. The off-normal events and potential upsets that could occur at each process step are described in a rectangle and given a name in a rectangle below the description, separated by a bold line. The name is used to track the event through the solution process.

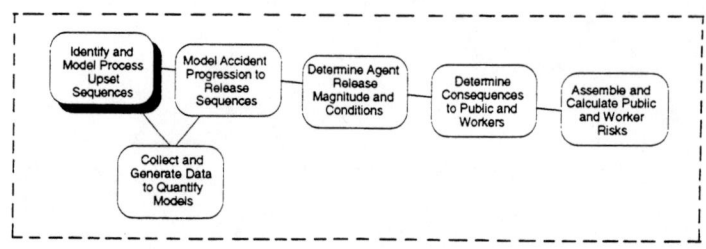

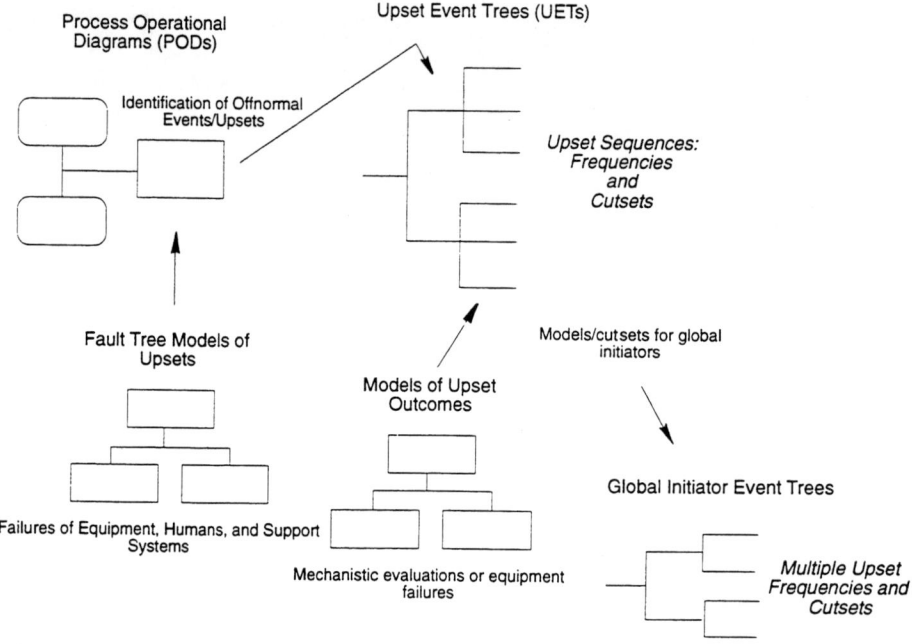

Figure 6-3. Identifying upsets. Source: U.S. Army, 1994e.

In the example of figure 6-4, the rockets in overpacks, called "pigs," are treated differently for some steps of the process than rockets not packed in pigs. An alternate path is indicated by a three-pronged connector and a new vertical path of rounded rectangles to the right of the diagram midline. In the case illustrated in figure 6-4, the alternative path tracks the pigs from Station 3 in the unpack area into the explosive containment vestibule by way of the bypass line. The empty on-site container ends up at the same point as containers from the main path in the process operational diagram (see U.S. Army, 1994e).

The steps in the QRA process are laid out in figure 6-5. The POD follows the plant process step by step to identify upset conditions. The QRA may quantify the likelihood of particular upset conditions with fault trees. Upset event trees define the initial outcomes of each upset in terms of the type of upset (e.g., spill, vapor leak, fire) and identify the location, agent, and quantity (i.e., the factors that define the release source term). Fault trees are used to quantify how likely various branches are. For initiating events that could cause multiple upsets, the analysts use global event trees to consider simultaneous releases. Accident progression event trees continue to track the upset to evaluate the effectiveness of barriers to release (e.g., mitigation systems such as HVAC filters) to determine public risk. Data analysis and systems analysis using fault trees follow standard techniques. Source term evaluation is automated through a set of "bins" according to rules that assign event tree sequences to source term "bins" that characterize the nature of the release based on information in the trees (e.g., the type and condition of munition, the agent, the location, the condition of barriers). Consequence analysis and external events (e.g., earthquakes, fires, tornadoes) are carried out in standard fashion. Details of all these modeling tools are provided in the methodology manual (U.S. Army, 1994e).

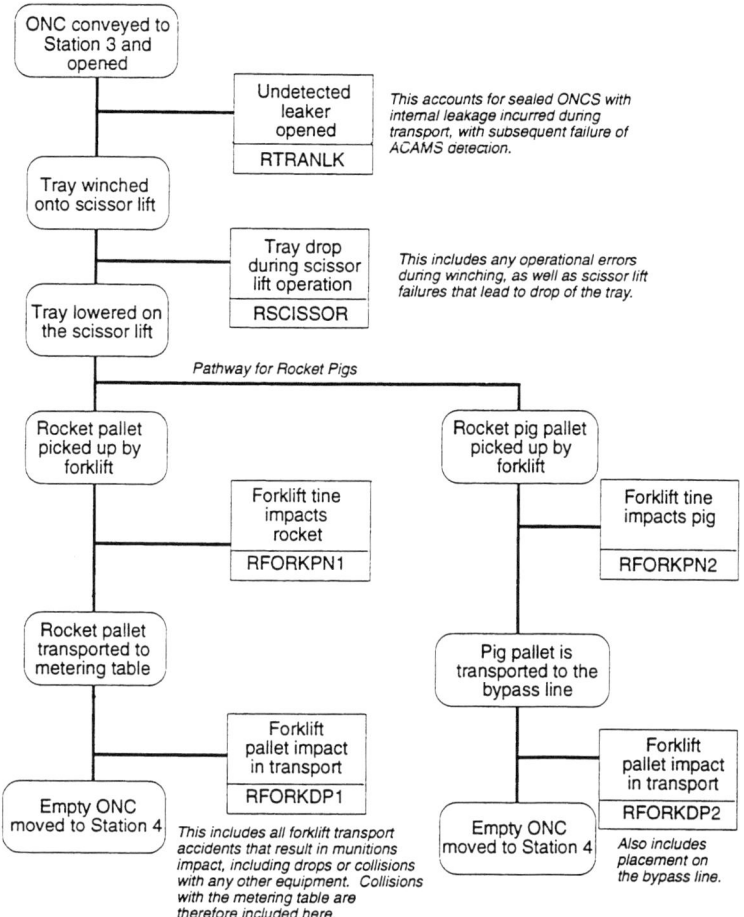

FIGURE 6-4 Sample portion of a rocket handling process operational diagram. Source: U.S. Army, 1994e.

Independent Review Committee Role and Evaluation

Detailed review of the accident QRA is being provided by a group of five experts (the Expert Panel, whose qualifications are discussed in an earlier section of this chapter) brought together under contract with MITRE and operating independently of project management. One member of the Stockpile Committee has been invited to monitor the Expert Panel review meetings and has been able to question individual analysts about details of the QRA.

The Expert Panel held its first meeting on November 1–2, 1994, at Edgewood, Maryland, and continued on November 3–4, 1994, at the TOCDF. The panel toured the Chemical Demilitarization Training Facility and TOCDF and were briefed on the demilitarization program and the QRA approach. At the TOCDF they met with the science advisor to the governor of Utah and the chair of the Utah Citizens Advisory Commission, as well as with site personnel and managers. Several members of the Expert Panel met with the QRA team at Edgewood on January 5–6, 1995, to review several technical aspects of the analysis in more detail.

The Stockpile Committee monitoring of the review process began with the second full meeting of the Expert Panel on February 1–2, 1995. The panel also met on March 16–17 and May 11–12. The Expert Panel selects a member to chair each meeting and prepares minutes on the meeting and a report summarizing comments and recommendations. The Army responds to each report in writing, explaining how they will address each comment. At later meetings, the Army presents and discusses work resulting from responses to the Expert Panel's suggestions. See table 6-1 for details of reports generated during the review process. Analysts performing each phase of

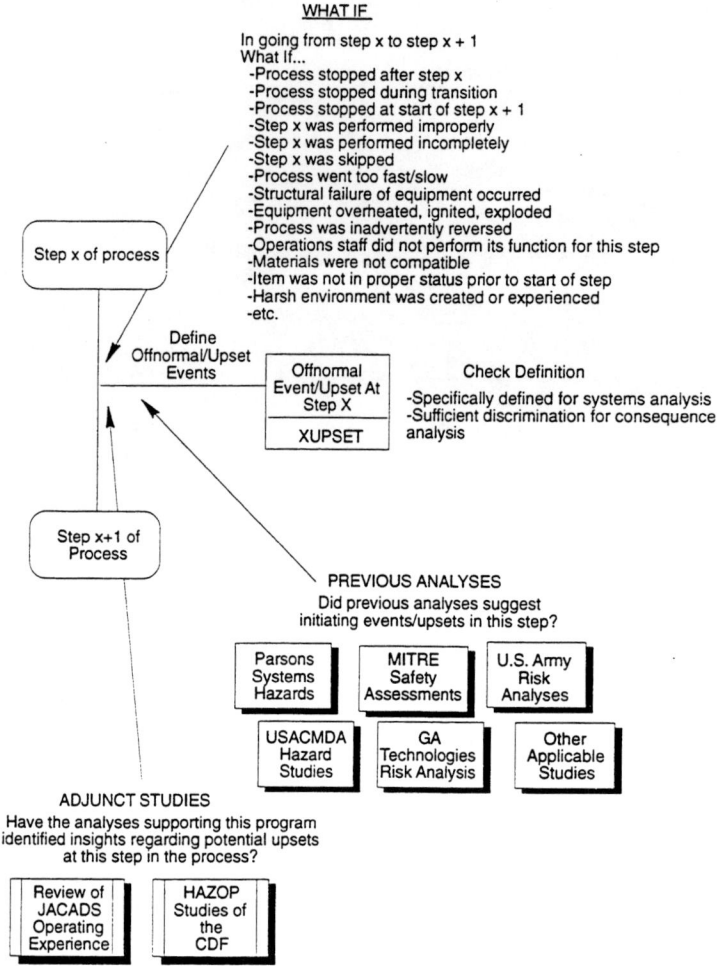

FIGURE 6-5 Schematic drawing of process operational diagram development. Source: U.S. Army, 1994e.

the work brief the panel. When meetings are held in Edgewood, specific analysts can attend the meeting when questions arise on their work. Table 6-2 details briefings to the Expert Panel. The panel also meets in executive session to decide how to carry out its functions and to discuss recommendations.

The Stockpile Committee finds that the review process is thorough and effective. The members of the panel are well qualified and are clearly independent. They probe deeply into the analysis and have received excellent cooperation from the Army and its contractor SAIC. The committee believes that the most valuable aspect of the review is the panel's focus on omissions rather than on the quality of the existing analysis. Unlike nuclear reactor quantitative risk assessments, which have been performed for many plants over many years, this is the first site-specific, full-scope quantitative risk assessment of a chemical disposal facility. Without the advantage of previous analyses and reviews, it seems more likely that there may be omissions. The search for what is not there is difficult, but the Expert Panel has the combined QRA experience and chemical process experience to provide a good second check on completeness of the analysis with respect to risk relevant scenarios.

Results

The second draft of the QRA for TOCDF campaigns 1 and 2[1] has been released and reviewed. The published version of the report will be released in the first quarter

---

[1] Tooele Chemical Agent Disposal Facility Quantitative Risk Assessment: Campaigns 1 and 2, SAIC-95/1006, prepared for U.S. Army Chemical Demilitarization and Remediation Activity, June 1995.

TABLE 6-1 Reports Associated with the Expert Panel Review of the Tooele Chemical Agent Disposal Facility Quantitative Risk Assessment

Report and Author(s)	Date
Quantitative TOCDF Risk Assessment, minutes of the Expert Panel meeting (S. Seth)	November 1–4, 1994
Draft recommendations, comments, and questions of the Expert Panel on TOCDF Risk Assessment (based on the first Expert Panel meeting, November 1–2, 1994) (G. Apostalakis, R.J. Budnitz, P.O. Hedman, G.W. Parry, R.W. Prugh)	January 15, 1995
Discussion of most significant comments from draft Expert Panel comments of the November meeting (SAIC)	February 1, 1995
Notes on the subcommittee meeting held at the SAIC office in Abingdon, January 4 and 5, 1995 (G. Parry)	January 12, 1995
Minutes of the second Expert Panel meeting, February 1–2, 1995 (S. Seth)	February 1995
Summary responses to comments from draft Expert Panel comments of the February meeting (SAIC)	March 17, 1995
Letter report on agent combustion and molten aluminum-agent reactions (P.O. Hedman)	March 21, 1995
Second report of the QRA Expert Panel (G. Apostalakis, R.J. Budnitz, P.O. Hedman, G.W. Parry, R.W. Prugh)	March 1995
TOCDF Risk Assessment: Minutes of the third Expert Panel meeting, March 16–17, 1995 (S. Seth)	March 1995
Third report of the QRA Expert Panel (G. Apostalakis, R.J. Budnitz, P.O. Hedman, G.W. Parry, R.W. Prugh)	April 1995
Comments from Meeting 2 (including clarifications from Meeting 3) (SAIC)	May 10, 1995
TOCDF Risk Assessment: Minutes of the fourth Expert Panel meeting, May 11–12, 1995 (S. Seth)	May 1995
Fourth report of the QRA Expert Panel (G. Apostalakis, R.J. Budnitz, P.O. Hedman, G.W. Parry, R.W. Prugh)	May 1995

of 1996 and is expected to be the same in substance as the draft report, with improved presentation.

The TOCDF QRA analyzes the acute and latent risks of both lethal and mild injuries to the public and site workers from accidental releases of agents. Thus the effects of normal operations like routine stack emissions are not included in this report. The impact of permitted stack releases will be evaluated in the health effects risk assessment that is being performed by the state of Utah. The current report is limited to the first two campaigns at TOCDF and includes risks associated with handling and transportation required for the two campaigns as well as processes within the TOCDF. The first campaign is a co-processing campaign, in which both M55 rockets and ton containers filled with nerve agent GB will be destroyed. The second campaign involves M55 rockets and spray tanks filled with nerve agent VX. All M55 rockets at Tooele will be destroyed during campaigns 1 and 2.

The purpose of the QRA for campaigns 1 and 2 is to support early use of risk insights for potential improvements prior to the start of the first two campaigns.

TABLE 6-2 Presentations to the Expert Panel Review of the Tooele Chemical Agent Disposal Facility Quantitative Risk Assessment

Organization	Presentation	Date
U.S. Army Risk and Surety Management Division		
	CSDP Risk Management	February 1–2, 1995
Science Applications International Corporation		
	TOCDF QRA Expert Panel Briefing	November 1–2, 1994
	Handouts for Agenda Item: • Preliminary Insights • Fires • Aircraft Impact • Seismic Analysis • Tornadoes • Transportation	February 1–2, 1995
	TOCDF Risk Management Plan	March 16, 1995
	Preliminary QRA Insights for the First Campaign	March 16, 1995
	Determination of Energetic Initiation and Leak Probabilities from Drops	March 16, 1995
	Aging Effects on Rockets Explosive Components	March 16, 1995
	Accident Progression Event Tree Update	May 12, 1995
	Uncertainty Analysis	May 12, 1995
	Update on Mechanistic Analysis Issues	May 12, 1995
	Filter Desorption	May 12, 1995
	Data Analysis and Synopsis of JACADS Experience	May 11–12, 1995

Significant safety improvements are being incorporated as a result of the QRA.

The QRA examines all potential causes of agent release during the two campaigns, including handling accidents (drops and impacts), mechanical failure, aircraft crashes, and severe environmental hazards such as earthquakes and fires. The report presents a number of other risk measures that are averages over selected parameters of the risk. On the basis of the expected (average) number of fatalities, comparisons show that the risk of the two campaigns is small compared to the risk of continued storage. When the full QRA is completed in 1996, more thorough and more meaningful comparisons will be available.

The results of the analysis are explained from various perspectives. One of the most useful identifies the sources of the highest frequency contributors to risk of lethalities (in terms of the expected number of lethalities). The effects of earthquakes dominate both the risk of processing and the risk of continued storage. Although severe seismic events are unlikely (4×10^{-5} per year, or once every 25,000 years), they are one of the few kinds of events or failures likely to cause significant release of agent. This is because agent is processed in very limited quantities at any given time. The risk to workers and the public combines the chance that an initiating event occurs with the chance that a significant release occurs given that event. Although the chance of a severe earthquake is low, it can cause significant damage and release. Other more likely events have little chance of causing significant release. The analysis shows that about 99 percent of the risk of processing comes from seismic events. Risk management activities now under way for TOCDF may further reduce the likelihood of release from seismic events and hence reduce overall risk from processing. Of the remaining 1 percent of the risk, about 94 percent is from accidents in handling in the storage yard. Because of substantial uncertainties in the modeling of handling accidents, additional work is being done to better characterize these events.

The current draft of the QRA for campaigns 1 and 2 presents results. The final report on the complete TOCDF QRA will present details of the calculations and models that led to these results. The bases for much of that information have been reviewed during the Expert Panel review sessions. At this writing, the analysis techniques appear to be thorough and properly applied, and the results appear to be reasonable in light of those methods. It is important that documentation of the analysis be completed in a timely fashion.

NRC EVALUATION OF THE ACCIDENT RISK PROGRAM

An important element for ensuring the quality of the QRA is the review by the Army's independent Expert Panel. The Stockpile Committee is satisfied that the Expert Panel is well constituted to perform its role and that meetings are held frequently enough and with sufficient time to provide adequate review. The committee is pleased to monitor the review meetings and have the opportunity to raise technical questions directly with the analysts performing the work. Because this QRA is the first of its kind, the committee would like to see the panel place even more emphasis on possible omissions from the assessment. The committee is especially pleased that the Army and the QRA team are responding thoroughly to issues raised by the Expert Panel.

The methods and depth of examination in the QRA appear to be appropriate. Follow-up with detailed engineering analysis and convening expert groups when difficult technical issues arise also appear to be thorough. The committee finds the current risk management process appropriate for resolving issues uncovered by the QRA. When a possible problem with agent-aluminum interaction was identified, the Army responded with additional engineering analysis to determine if the problem was the result of conservative modeling or was a real issue. When calculations could not eliminate concern, modifications to the operating plans and the system were developed to ensure that the chance of damage to the furnace and injury to nearby personnel would be reduced and would represent a very small contribution to total risk.

As discussed earlier, the committee has several concerns about the current constituent parts of the TOCDF risk assessment. Foremost among them is the concern that the studies are being performed as separate projects under diverse management with no obvious coordination. Accordingly, the committee believes that a single "TOCDF Risk Assessment Summary Report" should be published to present integrated results of the studies.

7

Findings and Recommendations

OVERVIEW

The Stockpile Committee has reviewed the status of the Tooele Chemical Agent Disposal Facility (TOCDF) at the Tooele Army Depot in Utah with respect to a number of recommendations made by the committee in three major reports and one letter report concerning improvements to be made to the Tooele facility prior to the start of agent operations. These reports are: *Evaluation of the Johnston Atoll Chemical Agent Disposal System Operational Verification Testing: Part II* (NRC, 1994a); *Review of Monitoring Activities Within the Army Chemical Stockpile Disposal Program* (NRC, 1994b); a letter report to the Assistant Secretary of the Army for Installations, Logistics and Environment to recommend specific actions to enhance further the Chemical Stockpile Disposal Program risk management process (NRC, 1993a); and *Recommendations for the Disposal of Chemical Agents and Munitions* (NRC, 1994c).

The following findings are based on the Stockpile Committee's evaluation of implementation of prior recommendations, on knowledge and observation of the baseline incineration system, on information provided by the Army and others, and on four site visits to the TOCDF—one in November 1991, one in March 1993, one in May 1994 shortly after the start of systemization, and one in March 1995 towards the end of the systemization period. Four subgroups of the committee also visited the TOCDF during the spring of 1995. These findings are based on information obtained prior to August 31, 1995, and before the completion of all the requirements that must be met by the Army before the start of or during the first year of agent operations.

FINDINGS

Responses to OVT II Report Recommendations

The following recommendations were made in the Stockpile Committee report *Evaluation of the Johnston Atoll Chemical Agent Disposal System Operational Verification Testing: Part II* (NRC, 1994a):

> Give safety considerations priority over production goals. (OVT2-1)

Current status. The Stockpile Committee reaffirms this position. The Army has stated its concurrence through a comprehensive safety oversight program. Nevertheless, the predominant focus of safety program implementation at the TOCDF seems to be upon agent-related operations and activities—with an apparent de-emphasis on nonagent operations. A strong safety culture in any organization requires consistent emphasis on *all* safety issues.

It is the committee's belief that the interrelationships and interdependencies between and among all on-site operations and activities mandate that agent- and non-agent-related safety programs should be on a par with each other prior to the start of agent operations.

The award fee criteria (or penalties) for EG&G Defense Materials, Inc., under their contract with the Army are still being finalized. The Stockpile Committee believes that the award fee criteria provide an opportunity for emphasizing safety priorities over production goals in the fee incentive (or penalty) structure.

> Proceed with Tooele systemization, and during systemization, conduct needed testing and improvement activities, including the following: (OVT2-2)
>
> • Develop and demonstrate an improved agent monitoring and identification system. (OVT2-2A)

Current status. A detailed discussion of this recommendation appears later in this chapter under responses to the *Monitoring* report recommendations.

> • Complete the brine reduction area (to include its pollution abatement system) performance tests, or develop a satisfactory brine disposal alternative. (OVT2-2B)

Current status. The brine reduction area, including its pollution abatement system, has been tested at both the Johnston Atoll Chemical Agent Disposal System (JACADS) and the TOCDF. It has passed the environmental compliance test at JACADS and systemization testing at the TOCDF.

- Demonstrate the dunnage furnace performance with various levels of chlorinated waste; if needed, either modify the pollution abatement system (e.g., add acid gas scrubbing) or limit feed materials to those that can be handled by the existing design; alternatively, satisfactory land disposal options must be identified. (OVT2-2C)

Current status. The dunnage furnace has been tested satisfactorily at JACADS and has passed two major nonagent systemization tests at the TOCDF. The dunnage furnace must still pass an RCRA-required trial burn test at the TOCDF after the start of agent operations.

- Review the probable levels of NO_x production from VX destruction and the allowable emission levels at the other continental U.S. sites requiring VX destruction; if appropriate, develop needed NO_x abatement systems. (OVT2-2D)

Current status. The TOCDF furnaces are expected to be able to meet NO_x emission requirements. No NO_x abatement systems are required.

- Develop and demonstrate the proposed hot-slag removal system for the liquid incinerator system. (OVT2-2E)

Current status. A hot-slag removal system has been installed on each liquid incinerator, but these have not been tested under hot-slag conditions because the Army decided there was no adequate surrogate test material for the highly variable slag to allow for complete testing of the system. The Army will first test the system as the slag accumulates from agent destruction operations.

- Eliminate furnace feed errors by improved monitoring and control of the deactivation furnace and metal parts furnace feed systems and by improved methods for tracking the various types of munitions. (OVT2-2F)

Current status. The Army retained MITRE Corporation to perform a detailed study of the furnace feed systems, including improvements made on the basis of experience at JACADS, and to recommend any further improvements. A detailed report (MITRE, 1994) describes the analysis of improvements that have been made in the munitions handling system and concludes that the changes minimize the chances of recurrence of the problems at JACADS. A second report (MITRE, 1995) describes further improvements in the system. The reconfigured munitions handling system is also being analyzed in detail in the risk assessment study being performed by Science Applications International Corporation (SAIC). The risk assessment should identify potential munitions handling problems that might lead to safety problems and provide an independent review of the systems involved.

- Address all problems associated with residual gelled mustard, in particular, the use of suited personnel to perform functions that were intended to be automated. (OVT2-2G)

Current status. The Army has developed an improved agent extraction verification system that avoids the need for manual intervention in routine operations.

Establish and maintain close working relationships with permitting agencies, and support these efforts with careful analysis of operating parameters to ensure that permits provide for safe destruction of agent, adherence to regulatory requirements, and effective plant operations. (OVT2-3)

Current status. The TOCDF has organized an Army-staffed program management unit focused on safety, quality assurance, and environmental oversight. A similar organization exists within the EG&G program management organization. Environmental oversight includes both permitting and environmental compliance. Close working relationships have been established with the Utah Division of Solid and Hazardous Waste (DSHW) to maintain responsive interactions to facilitate preparation of the RCRA permit as the TOCDF makes some final modifications based on systemization experience and other factors. Once agent operations start, the DSHW will maintain a presence on site during operations.

Establish programs, procedures, and management oversight to ensure continuing compliance with all environmental regulations. (OVT2-4)

Current status. EG&G has drafted an Environmental Compliance Plan (ENVCP), a document that summarizes general environmental compliance guidance, the background to the environmental impact statement, and the environmental requirements to which the TOCDF will adhere. Implementation of the ENVCP and strict adherence to the established environmental procedures should ensure continuing compliance with all environmental regulations.

Develop systems to improve overall management of safety. (OVT2-5)

Current status. The management structures in place at TOCDF and limited evidence gathered by Stockpile Committee members during site visits suggest that high quality safety management systems are in place. All action items identified in the Pre-Operational Survey must be resolved by the Army prior to the start of agent operations. The Stockpile Committee welcomes the Army's initiation of the Programmatic Lessons Learned program, creation of the Field Lessons Learned Review Team, commencement of Subject Area Reviews, and establishment of the Risk Management Plan, which is being developed by SAIC as an outgrowth of their TOCDF risk assessment work. The Stockpile Committee supports the Army's development of a process to identify "precursor events" at operating facilities (e.g., equipment failures and human errors) to maintain continuing emphasis on a philosophy of totally safe operations.

The Stockpile Committee is concerned about the focus on the start of agent operations at the TOCDF as the trigger for implementing full safety practices and about the lack of emphasis on industrial safety during the pre-agent, final systemization phase.

Complete the risk assessment for the Tooele Chemical Agent Disposal Facility during the systemization period. (OVT2-6)

Current status. Because experience from Operational Verification Testing (OVT) at JACADS resulted in a number of changes to the design and planned operations at the TOCDF, the committee wanted to ensure that the effects of those changes on risk would be examined. The Army has adopted this recommendation in stepwise fashion. Because of time constraints, the entire risk assessment will not be completed before the start of agent operations at the TOCDF. However, risk assessment pertinent to each campaign will be completed and lessons learned will be applied to the facility before the start of each campaign. The first analysis for campaigns 1 and 2 (GB and VX M55 rockets with co-processing of bulk items) was completed in April 1995. Modifications to the TOCDF and its operations are in progress and will be completed prior to the start of agent operations. The draft report documenting the analysis for the first two campaigns was completed on June 26, 1995. The complete TOCDF risk assessment is expected to be published in the first quarter of 1996. The committee believes that the Army's approach to the TOCDF risk assessment meets the intent of the committee's recommendation for the first two campaigns.

Finding 1. The Stockpile Committee finds that the Army has implemented or will soon implement the changes recommended in *Evaluation of the Johnson Atoll Chemical Agent Disposal System Operational Verification Testing: Part II.*

Responses to Monitoring Report Recommendations

The following recommendations were made as part of the Stockpile Committee's report *Review of Monitoring Activities Within the Army Chemical Stockpile Disposal Program*, (NRC, 1994b). The recommendations were grouped into three categories: general; agent/non-agent monitoring; and laboratory operations.

General Recommendations

The general recommendations were:

The Army should initiate a substantial program to upgrade the monitoring systems for continental U.S. sites. (MON-1)

The Army should obtain expert help at both the systems design and the equipment selection levels, perhaps by engaging a contractor with extensive experience in monitoring of trace species and in advanced instrument development. (MON-2)

The Army should undertake whatever instrument development is necessary to ensure that improved instrumentation is available to the chemical disposal program in suitably rugged and operational forms. (MON-3)

FINDINGS AND RECOMMENDATIONS

The Army should test and use new monitoring instrumentation at JACADS before such instrumentation is employed at Tooele. (MON-3)

The Army should plan to continually improve the monitoring system in areas where performance is presently limited by unavailability of suitable instrumentation. (MON-5)

Current status. Although existing instrumentation is adequate at the TOCDF, the Stockpile Committee continues to believe that the Army should investigate new monitoring technologies with the intent of continual improvement. The committee recognizes that the Army has instituted a program being conducted by the University of Denver (Fourier Transform Infrared technology) and SAIC (literature search and market survey of near real-time monitoring technologies) to evaluate improvements in monitoring technology. However, these efforts may be not be sufficient to ensure implementation of an advanced monitoring system.

Agent/Nonagent Monitoring Recommendations

The recommendations for agent/nonagent monitoring were:

Add the capability for positive identification of chemical agent species (chemical speciation) to the agent detection system and analytical laboratories at all of the disposal facilities in order to reduce the occurrence of false positives. (MON-6)

Current status. Speciation capability has been added to the laboratories at the Chemical Agent Munitions Disposal System (CAMDS), JACADS, and the TOCDF with new gas chromatograph/mass spectrometers. Although the Army is working on the problem of chemical speciation, the Automatic Continuous Air Monitoring System (ACAMS) does not yet have this capability.

Institute continuous monitoring for all agents present at each facility, including those in storage areas. (MON-7)

Current status. This requirement has been met by placing ACAMS monitors calibrated for various agents in the unpack areas of the TOCDF. A multiagent ACAMS is under development and being tested at CAMDS. Continuous monitoring has not yet been implemented in storage areas, but the Army has told the Stockpile Committee that plans are under way for continuous monitoring in these areas.

Reduce the time for confirmation of false positives. (MON-8)

Current status. This requirement has not been met. ACAMS alarms still require the laboratory analysis of Depot Area Air Monitoring System (DAAMS) samples to confirm false positives. A single false positive requires the shutdown of agent operations, but does not by itself initiate the response appropriate for a major agent release. False positive signals result in plant disruptions and the potential for increased human error and equipment degradation. The Army expects that the multi-agent ACAMS will have a lower false positive rate. (A dual detector ACAMS is expected to be ready for field tests by the end of 1995).

Evaluate the procedures for periodic testing of field sensors to ensure that false negatives are not possible if a significant release should occur. (MON-9)

Current status. The Army has developed a more comprehensive schedule for testing field sensors and has also installed some monitors that will more quickly and reliably signal detection of higher levels of agent in the event of a significant release.

Implement monitoring designed to provide more rapid response to high-level agent release. (MON-10)

Current status. Faster response monitors, set to higher detection levels (immediately dangerous to life and health), have been installed in the unpack area of the facility. A new type of detector (based on ion mobility), which has about a 30-second response time, is presently under development.

Evaluate the benefits of more frequent analysis of facility stack gases for nonagent trace contaminants. (MON-11)

Current status. The TOCDF will be undergoing a series of RCRA- and TSCA-required state-supervised trial burns for the first 24 months of operations; a minimum of 16 trial burns (each consisting of 3 or 4 runs) are scheduled. The results of the trial burns will be used to revise the health risk assessment and formulate a monitoring approach for products of incomplete combustion, particulates, heavy metals, halogenated dioxins and furans, and volatile and semivolatile organics.

Laboratory Operations Recommendations

The recommendations for laboratory operations were:

Increase the automation of sample handling and laboratory operations to ensure better quality control and efficiency. (MON-12)

Current status. The TOCDF laboratory incorporates a new bar code system for identifying samples and an improved automated record system. The Stockpile Committee believes these innovations constitute a major improvement in quality control over the procedures at JACADS. The committee still believes that a system is needed to track and evaluate the cause of lab errors. This system will continue to improve the quality of laboratory operations.

Give laboratory personnel a variety of tasks that ensure optimal attention and performance. (MON-13)

Current status. The laboratory now handles nonagent analysis of flue gas samples and includes mass spectroscopic analysis for agent in addition to gas chromatography. Laboratory operators are not presently cross-trained for all these techniques. The Army plans to start cross-training in the next few months to include instrument training, hazardous waste training, testing new monitors, and statistical analysis of data. Cross-training is an excellent way to enhance personnel attention and individual operator job performance.

Give blind challenges to the laboratory. (MON-14)

Current status. The Army has established and implemented a system for providing double blind challenges to the laboratory in a way that will provide frequent checks on the quality and reliability of normal laboratory operations

Perform a detailed error analysis of the laboratory system and procedures. (MON-15)

Current status. The Army plans to have such a capability in place late in 1995.

Finding 2. The Stockpile Committee finds that the Army has implemented or will soon implement most of the changes recommended in *Review of Monitoring Activities Within the Army Chemical Stockpile Disposal Program.*

Responses to Risk Letter Report Recommendations

Recommendations for Facilities in the Continental United States

The Stockpile Committee, in a letter report to the Assistant Secretary of the Army for Installations, Logistics and Environment, on January 8, 1993 (NRC, 1993a), made the following recommendations:

A site-specific, full-scope, scenario-based risk assessment should be performed for each continental U.S. facility, starting with the Tooele facility. (RISK-1)

Current status. SAIC completed the first phase of an accident quantitative risk assessment (QRA) for the TOCDF, and a draft report summarizing the results was circulated to the committee on June 26, 1995 (SAIC, 1995b). The first phase covers the portions of the facility that will be involved in the first two campaigns (GB and VX M55 rockets with co-processing of bulk items). The full accident QRA for the storage risks and all campaigns will be completed in early 1996. For technical and administrative reasons, there are multiple risk assessment projects and reports: an accident quantitative risk assessment, a health effects risk assessment, and a reconfiguration risk assessment. Organizations other than SAIC may perform some of these studies. This situation could lead to misunderstanding among reviewers, government agencies, and the public. It is essential that the Army adopt standard language recognizing the ensemble of risk-related projects as "the risk assessment." The individual studies should always be referred to as elements of the wider risk assessment.

Each site-specific risk assessment should include the case of continued storage without disposal as one scenario. (RISK-2)

Current status. The current status of the TOCDF risk assessment limits analysis to activities associated with the first and second campaigns. Risk associated with continued storage will be addressed later. The accident QRA team is currently obtaining additional expert assistance to help address specific aspects of the risk of continued storage and associated handling.

FINDINGS AND RECOMMENDATIONS

The risk assessments should be quantitative and include the following features:

- bottom-line results on the health effects to on-site and off-site personnel in terms of likelihood and consequence, including a site-specific atmospheric and health effects analysis and an analysis of emergency response capability;
- a clearly defined set of scenarios that, taken together, provide a comprehensive representation of the risk;
- dependency matrices that display inter- and intra-system dependencies;
- a human action analysis that represents the human role in controlling risk;
- quantification of risk from all causes, including both internal events (plant and plant-people failures) and external events (earthquakes, fires, floods, aircraft crashes, etc.);
- site-specific hardware, software, procedures, training programs, maintenance practices, and operations personnel (including site-specific storage facilities and munitions handling activities);
- risk contributors in such terms as random failures, common cause failures, multiple failures, and human error; and
- an uncertainty analysis to display clearly how much confidence the analysts have in the precision of the quantitative results. (RISK-3)

Current status. The Stockpile Committee believes that the accident quantitative risk analysis is progressing toward meeting the goals of this recommendation.

Broad Guidelines for Risk Management

The Stockpile Committee provided five broad guidelines for conducting the site-specific risk analyses:

Modern, up-to-date methodologies should be employed, such as those found in the risk assessments reported in NUREG-1150. (RISK-4)

Current status. The accident QRA meets this guideline.

The risk assessments should be conducted by organizations with recognized expertise in the field, but not otherwise involved in the CSDP. In a similar vein, independent peer reviews are an absolute requirement. (RISK-5)

Current status. SAIC has recognized capability for performing risk assessments but is involved in the Chemical Stockpile Disposal Program (CSDP) because it has been a long-term support contractor for the Army. The Army reports that SAIC was chosen to avoid a long procurement process for selecting an independent contractor that would have delayed the program. The Army retained another long-term support contractor, MITRE Corporation, to organize an expert panel to oversee the SAIC risk assessment activity. Five experts were chosen for the oversight panel (Expert Panel) and the Stockpile Committee has reviewed their qualifications and concluded that they are all well qualified. One of the members is a combustion expert from Brigham Young University in Salt Lake City, who provides some degree of local perspective. A member of the Stockpile Committee with expertise in risk assessment was invited to monitor the meetings of the Expert Panel. The Stockpile Committee believes that the Army has complied with the intent of this guideline.

Local representatives of neighboring communities must be involved early. Their concerns about the CSDP may be substantial, and will warrant consideration throughout the analysis process. (RISK-6)

Current status. The Army attempted to involve the community as early as the spring of 1994 and at intervals since then. However, key elements of the community seem to have missed or misinterpreted these interactions. Additional efforts are required if the involvement is to be successful. The lessons learned here could be applied at future sites.

Emphasis must be placed on human reliability factors, particularly in light of the human factors issues raised by the Stockpile Committee in reviewing the first phase of Operational Verification Testing at JACADS. (RISK-7)

Current status. The QRA team has brought human reliability experts into the project. However, to this point, because of the limited quantity of agent that can be processed in the facility, no human-caused events significant to off-site release have been identified. The committee expects that the impact of human reliability on worker risk and stockpile handling accidents could be important. The committee believes that SAIC should examine these areas.

To avoid overstatement of the results it is important that the confidence levels of the risk parameters be fully displayed. It is this process of quantifying the uncertainty in the risk that will establish the reliability of the conclusions. Experience has indicated that the results of a risk assessment provide valuable information on the importance of different contributors to risk, not only in terms of hardware failures but also in terms of human errors and deficiencies in procedures and software. Thus the risk assessment can lead to process changes that reduce overall risk. (RISK-8)

Current status. The Expert Panel is urging the QRA team to expand and clarify their efforts to characterize uncertainty. The Stockpile Committee expects that the QRA will meet the intent of this recommendation and will examine this area as the work proceeds. The committee is aware that several risk issues were raised as a result of the detailed risk assessment and that changes to equipment and procedures were made where appropriate to mitigate specific risks.

Finding 3. The Stockpile Committee finds that the Army has implemented or will soon implement the analyses and actions recommended in the risk letter report.

Responses to the Recommendations Report

The Stockpile Committee report, *Recommendations for the Disposal of Chemical Agents and Munitions* (NRC, 1994c), presented a number of programmatic recommendations that pertain, in part, to the Tooele Chemical Agent Disposal Facility. The report addressed the potential for and made recommendations regarding alternative technologies to the baseline system. Because the baseline system was already under construction for the TOCDF, those recommendations did not pertain to that facility.

Minimize Total Risk

The Chemical Stockpile Disposal Program should proceed expeditiously and with technology that will minimize total risk to the public at each site. (REC-1)

Current status. The TOCDF incorporates the baseline incineration system. Improvements have been made to reduce risk over the prototype baseline design in operation at JACADS. Considerable safety testing and configuration certification have been performed during systemization to meet operating permit conditions. Because more than 40 percent of the stockpile is at the TOCDF, safe operation of the facility is a major element in the Army's mission for expeditious chemical stockpile disposal.

Assessment of Risk

Recommendations 2 through 5 of the *Recommendations* report pertain to risk assessment. However, all are not applicable to the TOCDF. The Stockpile Committee recommended that the Army's risk analyses completed in 1987 as input to the Programmatic Environmental Impact Statement (U.S. Army, 1988) be *updated* as soon as possible, on *a site-specific basis*. In addition, the Stockpile Committee recommended that the new risk analyses should *explicitly account for latent health risks from storage, handling, and disposal*. The Stockpile Committee noted that these *analyses will confirm or refute the present belief that maximum safety dictates prompt disposal.*

Current status. Work on the risk assessment is not yet complete, but a seismic analysis of the TOCDF strongly implies that storage area impacts would be even more severe than disposal system impacts than previously thought, and that maximum safety at the TOCDF dictates prompt disposal. The analytical work for the accident QRA of the first two campaigns at the TOCDF was completed as of late June 1995. Several aspects of the work on the risk of continued storage and stockpile reconfiguration and handling are continuing or will be revised. This work is not significant to the risk associated with the first two campaigns. The Stockpile Committee strongly believes that a final overview report should be prepared that presents a clear and integrated view of all potential risks and states the estimated confidence levels associated with the results. Further, the Stockpile Committee supports the Army's plan to keep the TOCDF risk assessment current as changes are made to the facility in the future and to integrate the assessment with a TOCDF Risk Management Plan.

Public Concerns

The sixth recommendation focused on public concerns:

FINDINGS AND RECOMMENDATIONS

The Army should develop a program of increased scope aimed at improving communication with the public at the storage sites. In addition, the Army should proactively seek out greater community involvement in decisions regarding the technology selection process, oversight of operations, and plans for decommissioning facilities. Finally, the Army should work closely with the Chemical Demilitarization Citizens Advisory Commissions, which have been (or will be) established in affected states. There must be a firmer and more visible commitment to engaging the public and addressing its concerns in the program. (REC-6)

Current status. The committee finds the Army's efforts in Utah to obtain community input into the risk assessments were substantial, but not especially productive. The committee believes it is essential to obtain such involvement prior to the beginning of a risk assessment, as well as during its implementation, to improve risk communication and, ultimately, to gain public acceptance of the results.

The committee finds that the Army has begun to implement a large and comprehensive public information program in Tooele. This program is particularly impressive because few resources had been devoted to public information and outreach until recently. Nevertheless, the list of activities either planned or under way in this program suggests that not enough attention may have been given to soliciting citizen input into programmatic decisions.

Because emergency management issues are likely to become more important to the public when the TOCDF starts operations, efforts should be made by the Army to include the public in the Chemical Stockpile Emergency Preparedness Program (CSEPP) and to integrate more fully the Army's public outreach with the CSEPP.

The Occupational Safety and Health Administration recently approved personal protective equipment for responders. The committee is aware that, to both the Utah Division of Comprehensive Emergency Management (CEM) and the Tooele County Department of Emergency Management, the issues of funding a core team of responders and conducting associated training of personnel in Utah are key roadblocks to implementation of the CSEPP.

The risks presented by the stockpile require that emergency response plans be completed and exercised to ensure preparedness and successful response in the event of an actual release. The Utah CEM has indicated that, because of the lack of national planning standards, guidance has been incomplete with regard to reentry, emergency medical services, and recovery phase operations, and that this in turn has led to less than effective training and exercises (Utah CSEPP Readiness Issues, presentation by the Utah CEM to the Stockpile Committee, March 29, 1995). Moreover, some counties opted not to participate in exercises, raising concerns about their level of preparedness for an emergency. The lack of national planning standards cannot be permitted to interfere with important planning for public safety and, if necessary, the Army must step in to rectify the situation.

The committee finds that the local CSEPP emergency planning efforts are not complete, as evidenced by the appendices of the Tooele County Emergency Operations Plan, which are still in a draft version.

The committee finds that the Communications Plan for Tooele County and the planned implementation of the communications system linking important operations centers in the emergency planning zone are not yet complete. In addition, as of this writing, public notification tone alert radios were not yet in place.

Baseline Incineration System

Recommendations 7 through 13 applied to the baseline incineration system. Recommendation 12 was:

The Chemical Stockpile Disposal Program should continue on schedule with implementation of the baseline system, unless and until alternatives are developed and proven to offer safer, less costly, or more rapidly implementable technologies (without sacrifice in any of these areas). Baseline system improvements should be implemented as identified and successfully demonstrated. (REC-12)

Current status. The Programmatic Lessons Learned and Subject Area Review programs are good vehicles for identifying baseline system improvements and implementing them as appropriate across disposal facilities at different sites.

Carbon Filters

Recommendation 13 addressed the possibility of adding activated carbon filter beds to the exhaust gas stack:

The application of activated charcoal filter beds to the discharge from baseline system incinerators should be evaluated in detail, including estimations of the magnitude and consequences of upsets, and site-specific estimates of benefits and risks. If warranted, in terms of site-specific advantages, such equipment should be installed. (REC-13)

Current status. The Army is presently doing the recommended evaluation. Filters can be added to the TOCDF facility at a later date if they provide a favorable benefit-to-risk ratio. Carbon filter bed testing and further evaluation are tentatively planned at the TOCDF, if the initial evaluations are positive.

Alternative Technologies and Stockpile Stability

Recommendations 14 through 20 addressed issues of alternative technologies and stockpile stability—not directly pertinent to the TOCDF systemization.

Program Staffing

The final recommendation pertained to the entire program and is related to concerns about the safety implications of adequate staffing. With the finding that, as the program expands in scope, a shortage of skilled staff could potentially compromise safety and have obvious implications for slowing the program down, with attendant increased risks, the Stockpile Committee recommended:

> The Army should establish a program to incrementally hire (or assign military) personnel to ensure that staff growth is consistent with the workload and with technical and operational challenges. These additional personnel must be assigned and trained before the project office gets deeply involved in addressing each challenge. (REC-21)

Current status. The Stockpile Committee has noted the addition of qualified personnel, both in the office of the Program Manager for Chemical Demilitarization (PMCD) and in contractor organizations at TOCDF.

The committee has some concerns that major Army leadership is new and that acting positions have not been permanently filled. Additional PMCD staffing may be needed as activities expand in the future. The present level of staffing at TOCDF appears appropriate for safe and environmentally compliant operation of the facility. The retirement of the TOCDF EG&G general manager just before to the start of agent operations was a matter of concern. It appears that prompt action was taken to manage the transition safely.

Finding 4. The Stockpile Committee finds that the Army has implemented or will soon implement changes pertinent to the TOCDF that were recommended in the *Recommendations* report.

RECOMMENDATIONS

Based on the Stockpile Committee's evaluation of the status of the TOCDF with respect to recommendations made in previous reports, the committee is generally satisfied with the progress made and recommends the following actions pertaining to safety and performance be taken at the TOCDF:

Duration of TOCDF Operations

Recommendation 1. Development and implementation of the overall safety program at the TOCDF must be given high priority.

Recommendation 2. Safety and environmental performance goals should be given at least equal weight with production goals in establishing award fee criteria.

Recommendation 3. Applicable portions of the accident quantitative risk assessments must be completed and all safety-related concerns resolved before the start of specific agent-destruction campaigns.

Recommendation 4. A substantial effort should be made by the Army to enhance interactive communications with the host community and the Utah State Citizens Advisory Commission on issues of mutual concern (e.g., various elements of the Chemical Stockpile Emergency Preparedness Program, decontamination and decommissioning, future use of the facility, and risk reduction).

Coordinated with the Start of Agent Operations

Recommendation 5. The Army should increase efforts to work with the Utah Division of Comprehensive Emergency Management to ensure that first-responders have been adequately trained to

FINDINGS AND RECOMMENDATIONS

use the personal protective equipment approved by the Occupational Safety and Health Administration. Tooele County must ensure their capability for responding to an emergency, especially because this condition relates to state requirements for the start of agent operations.

Recommendation 6. The Army, and where appropriate the Federal Emergency Management Agency (FEMA), should ensure that local and state Chemical Stockpile Emergency Preparedness Program plans for responding to potential chemical events are complete and well exercised as soon as possible.

Recommendation 7. The Army/FEMA should provide the necessary resources for completing the communications system planned by the Tooele County Department of Emergency Management.

Prior to the Start of Agent Operations

Recommendation 8. All mandatory requirements of the Army's Pre-Operational Survey must be satisfied.

Recommendation 9. The liquid incinerator and deactivation furnace system must have demonstrated a destruction removal efficiency of 99.9999 percent (6-nines) during surrogate trial burns.

Recommendation 10. High-quality, adequately staffed safety management systems must be completely implemented (including procedures for testing critical equipment; all necessary operating, maintenance, and emergency procedures; management of change procedures; training and cross-training programs; programmatic lessons-learned activities; subject area reviews; and other safety oversight activities).

During the First Year of Agent Operations

Recommendation 11. The liquid incinerator and the deactivation furnace system must pass all required Resource Conservation and Recovery Act trial burns; and the deactivation furnace system must also pass required Toxic Substances Control Act trial burns.

Recommendation 12. Testing and certification of the brine reduction area and the dunnage incinerator should be completed at the TOCDF, or a satisfactory disposal alternative must be implemented.

Recommendation 13. Performance of the slag removal system for the liquid incinerators should be demonstrated when sufficient slag has accumulated.

Recommendation 14. The Risk Management Plan must be fully implemented.

Recommendation 15. A comprehensive, integrated, and clear TOCDF risk assessment study, including a full description of all significant acute and latent agent and nonagent risks associated with disposal operations, as well as with the continued maintenance of the Tooele chemical stockpile, should be completed. A full explanation of the uncertainties associated with the various estimates should be included.

Recommendation 16. A system for documenting and tracking unexpected upsets, errors, failures, and other sources of problems that have led to "near misses" during operation of the facility should be developed as soon as possible. A program for integrating this information into a plan for continual safety improvements at the TOCDF should be implemented.

Recommendation 17. An active program for continual improvement of monitoring instrumentation, including techniques for more rapid recognition of significant levels of agent release, should be pursued.

Recommendation 18. Evaluations of the stack-gas carbon filter bed system should be continued.

Appendices

Appendix A

Public Law 102-484—Oct. 23, 1992
(Extract)

Subtitle G—Chemical Demilitarization Program

SEC. 171. CHANGE IN CHEMICAL WEAPONS STOCKPILE ELIMINATION DEADLINE

Section 1412(b)(5) of the Department of Defense Authorization Act, 1986 (50 U.S.C. 1521 (b)(5)), is amended by striking out "July 31, 1999" and inserting in lieu thereof "December 31, 2004."

SEC. 172. CHEMICAL DEMILITARIZATION CITIZENS ADVISORY COMMISSIONS

(a) ESTABLISHMENT.—(1) The Secretary of the Army shall establish a citizens' commission for each State in which there is a low-volume site (as defined in section 180). Each such commission shall be known as the "Chemical Demilitarization Citizens' Advisory Commission" for that State.

(2) The Secretary shall also establish a Chemical Demilitarization Citizens' Advisory Commission for any State in which there is located a chemical weapons storage site other than a low-volume site, if the establishment of such a commission for such State is requested by the Governor of that State.

(b) FUNCTIONS.—The Secretary of the Army shall provide for a representative from the Office of the Assistant Secretary of the Army (Installations, Logistics, and Environment) to meet with each commission under this section to receive citizen and State concerns regarding the ongoing program of the Army for the disposal of the lethal chemical agents and munitions in the stockpile referred to in section 1412(a)(1) of the Department of Defense Authorization Act, 1986 (50 U.S.C. 1521(a)(1)) at each of the sites with respect to which a commission is established pursuant to subsection (a).

(c) MEMBERSHIP.—(1) Each commission established for a State pursuant to subsection (a) shall be composed of nine members appointed by the Governor of the State. Seven of such members shall be citizens from the local affected areas in the State; the other two shall be representatives of State government who have direct responsibilities related to the chemical demilitarization program.

(2) For purposes of paragraph (1), affected areas are those areas located within a 50-mile radius of a chemical weapons storage site.

(d) CONFLICTS OF INTEREST.—For a period of five years after the termination of any commission, no corporation, partnership, or other organization in which a member of that commission, a spouse of a member of that commission, or a natural or adopted child of a member of that commission has an ownership interest may be awarded—

(1) a contract related to the disposal of lethal chemical agents or munitions in the stockpile referred to in section 1412(a)(1) of the Department of Defense Authorization Act, 1986 (50 U.S.C. 1521(a)(1)); or

(2) a subcontract under such a contract.

(e) CHAIRMAN.—The members of each commission shall designate the chairman of the commission from among the members of the commission.

(f) MEETINGS.—Each commission shall meet with a representative from the Office of the Assistant Secretary of the Army (Installations, Logistics, and Environment) upon joint agreement between the chairman of the commission and that representative. The two parties shall meet not less often than twice a year and may meet more often at their discretion.

(g) PAY AND EXPENSES.—Members of each commission shall receive no pay or compensation for their involvement in their activities of the commission.

(h) TERMINATION OF COMMISSIONS.—Each commission shall be terminated after the stockpile located in that commission's State has been destroyed.

Appendix B

Chemical Stockpile Disposal Program

THE CALL FOR DISPOSAL

The United States has maintained a stockpile of highly toxic chemical agents and munitions for more than half a century. Three unitary[1] agents are stored and exist largely as liquids: nerve agent VX, a high-boiling point liquid that will adhere to surfaces for days or weeks; nerve agent GB (sarin), a liquid that evaporates quickly and has a volatility similar to water; and mustard, a blister agent that evaporates slowly. These agents are stored in a variety of munitions and containers.

Lethal chemical agents are extremely hazardous, which is why they have been used in weapons. The manufacture of such agents and munitions and their subsequent stockpiling were undertaken in the belief that they were valuable as deterrents to similar materials being used against U.S. forces. That deterrence is no longer considered necessary. Consequently, the United States can no longer justify the continuing risk and expense of storage.

In an attempt to avoid the worldwide risk posed by chemical warfare, the United States is entering into agreement with many other nations to rid the world of all such materials. There is ample incentive for disposing of U.S. chemical agents and munitions as promptly as safe procedures permit.

In 1985, Congress passed Public Law 99-145 initiating the Chemical Stockpile Disposal Program (CSDP) to eliminate the unitary chemical stockpile, starting with an "expedited" effort to dispose of M55 rockets, a particularly hazardous munition. The program was expanded to treat the entire stockpile and led to the development of the current baseline incineration system. In 1992, after setting several intermediate goals and dates, Congress enacted Public Law 102-484 directing the Army to dispose of the entire unitary chemical warfare agent and munitions stockpile by December 31, 2004.

DISPOSAL PROGRAM BACKGROUND AND ROLE OF THE NATIONAL RESEARCH COUNCIL

The Army's search for the best disposal system for bulk agents and munitions has continued for some time, with input from several committees of the National Research Council. Prior to 1969, disposal was mainly by land burial, open pit burning, and deep ocean dumping.[2] An NRC review committee (NAS, 1969) concluded that:

> It should be assumed that all agents and munitions will require eventual disposal and that dumping at sea should be avoided. Therefore, a systematic study of optimal methods of disposal on appropriate military installations, involving no hazards to the general population and no pollution of the environment, should be undertaken.

The use of the terms "*no* hazard" and "*no* pollution" is unfortunate. The stockpile *is* a hazard, and both storage and disposal entail some risk. The only way to eliminate the hazard and associated storage risk is to eliminate the materials themselves.

The Army commissioned studies of different disposal technologies and tested several in the 1970s, including incineration and chemical neutralization (Moynihan et al., 1983). In 1982, the Army selected component disassembly and incineration with

---

[1] The term unitary distinguishes a single chemical loaded in munitions or stored as a lethal material. More recently, binary munitions have been produced in which two relatively safe chemicals are loaded in separate compartments to be mixed to form a lethal agent after the munition is fired or released. The components of binary munitions are stockpiled apart, in separate states. They are not included in the present Chemical Stockpile Disposal Program. However, under the Chemical Weapons Convention of 1993, they are included in the munitions that will be destroyed.

[2] Dumping at sea was later banned by the Marine Protection, Research, and Sanctuaries Act of 1972 (P.L. 92–532).

associated pollution abatement systems, now known as the baseline system, as the preferred disposal system.

The NRC Committee on Demilitarizing Chemical Munitions and Agents was formed in August 1983 to review the status of the stockpile and technologies for disposal. That committee reviewed a range of technologies and, in its final report in 1984, endorsed incineration as an adequate technology for the safe disposal of chemical agents and munitions (NRC, 1984). The committee also concluded that the stockpile was well maintained and posed no imminent danger but added, "It is not possible to give assurance at this time that an increased rate of deterioration may not occur within the relatively near future."

In 1987, at the request of the Undersecretary of the Army, the Committee on Review and Evaluation of the Army Chemical Stockpile Disposal Program (referred to as the Stockpile Committee) was established under the aegis of the National Research Council Board on Army Science and Technology to provide the Army with technical advice and counsel on specific aspects of the disposal program. Under this charter, the Army has requested and received 14 reports from the Stockpile Committee.

Construction of the Johnston Atoll Chemical Agent Disposal System (JACADS), the first facility to bring together and integrate the elements of the baseline system, was begun in 1984. JACADS began operations using agents in July 1990 with Operational Verification Testing (OVT) that concluded in March 1993. The MITRE Corporation was engaged to monitor four test series (MITRE, 1991, 1992, 1993a, 1993b) and to provide a summary report upon conclusion of OVT (MITRE, 1993c). The Stockpile Committee issued a preliminary review and commentary on MITRE's reports in July 1993, *Evaluation of the Johnston Atoll Chemical Agent Disposal System Operational Verification Testing: Part I* (NRC, 1993b), including comments and broad recommendations on the implications of JACADS performance for disposal facilities in the continental United States. The committee then issued a more detailed review containing expanded recommendations for improvement of the baseline system, *Evaluation of the Johnston Atoll Chemical Agent Disposal System Operational Verification Testing: Part II* (NRC, 1994a).

In 1989, construction of the first disposal facility in the continental United States, the Tooele Chemical Agent Disposal Facility (TOCDF), was begun at the Tooele Army Depot in Utah. The design of the TOCDF represents a second generation baseline system, incorporating improvements based on experience with the JACADS facility, advances in technology, and recommendations made by the Stockpile Committee. Pre-operational testing, or "systemization," of the TOCDF started in August 1993.

During the systemization period, additional modifications were made to systems and procedures at the TOCDF in response to recommendations by the Stockpile Committee in the two OVT reports mentioned above and in *Review of Monitoring Activities Within the Army Chemical Stockpile Disposal Program* (NRC, 1994b) and *Recommendations for the Disposal of Chemical Agents and Munitions* (NRC, 1994c).

In addition, the Stockpile Committee issued a letter report concerning the chemical stockpile disposal risk management process (NRC, 1993a). In that report, the committee recommended that a site-specific risk assessment be performed at each continental U.S. site prior to the start of agent operations. Each risk assessment is expected to include all site operations, including continuing risks from storage as well as risks from accidental agent releases and from chronic exposures during plant operations. The Army has retained Science Applications International Corporation (SAIC) to perform the site-specific risk assessments.

DESCRIPTION OF THE STOCKPILE

Agents

The two principal types of agent in the U.S. stockpile are nerve agents (GB and VX)[3] and blister or mustard agents (H, HD, HT). Each is found in a variety of containers and munitions.

Nerve agents are organophosphonate compounds that contain phosphorus double-bonded to an oxygen atom and single-bonded to a carbon atom. Nerve agents are highly toxic and lethal in both liquid and vapor forms. In pure form, the nerve agents are practically colorless and odorless. GB evaporates at about the same rate as water and is relatively nonpersistent in the environment. VX evaporates much more slowly and can persist for a long time under average weather conditions.

---

[3] GB is O-isopropyl methylphosphonofluoridate. VX is O-ethyl, S[2-(diisopropyl amino)ethyl]methylphosphonothiolate.

Bis(2-chloroethyl)sulfide is the principal active ingredient in blister agents, or mustard.[4] Mustard has a garlic-like odor. It presents both vapor and contact hazards. Because it is practically insoluble in water, mustard is very persistent in the environment and can contaminate soils and surfaces for a long time.

Containers and Munitions

The stockpile of unitary chemical agents can be found in containers (various bombs stored without explosives, aerial spray tanks, and ton containers) and munitions (land mines, M55 rockets, bombs, artillery projectiles, and mortar projectiles) (see figures B-1, B-2, and B-3). Some munitions are stored with no explosives or propellant, whereas others contain some combination of fuse, booster, burster, and propellant (table B-1). These components are referred to collectively as "energetics." They include a variety of chemical compounds that must be eliminated as part of the chemical stockpile disposal operation.

The fuse, a small, highly sensitive explosive element, initiates an explosive chain by detonating a booster. The booster is an intermediate charge sensitive enough to be detonated by the fuse and energetic enough to detonate the much larger burster. The burster, the end of the chain, bursts the munition with sufficient energy to disperse the agent. The M55 rocket also contains an integral solid rocket propellant that can be removed only by cutting open the rocket.[5]

---

[4] Names such as mustard gas, sulfur mustard, and yperite have also been applied to this agent. The term mustard "gas" is often used, but the chemical is a liquid at ambient temperature.

[5] Fuses may contain cyclonite, lead styphnate, lead oxide, barium nitrate, antimony sulfide, tetracine, and potassium chlorate. Bursters may have tetryl, tetrytol (tetryl plus trinitrotoluene [TNT]), or Composition B (cyclonite plus TNT). Propellants may include nitrocellulose, nitroglycerine, lead stearate, triacetin, dibutylphthalate, and 2-nitro diphenylamine.

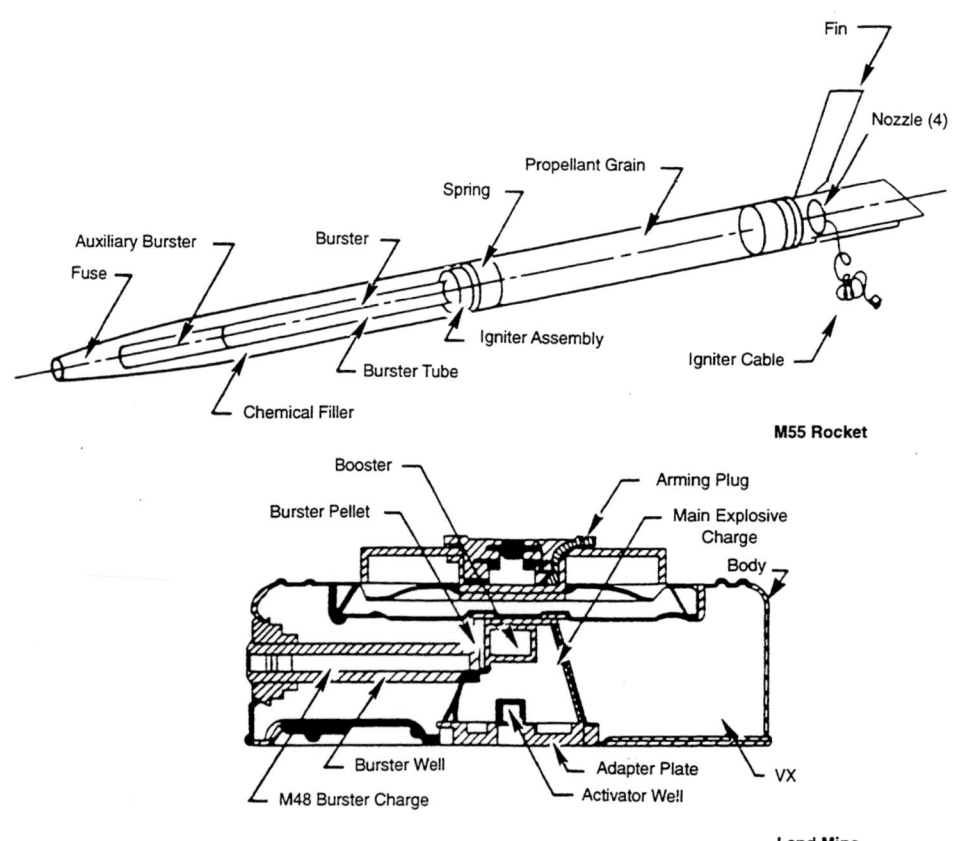

FIGURE B-1 M55 rocket and M23 land mine. Source: USATHAMA, 1982, 1983; NRC, 1993c, 1994a,c.

APPENDIX B

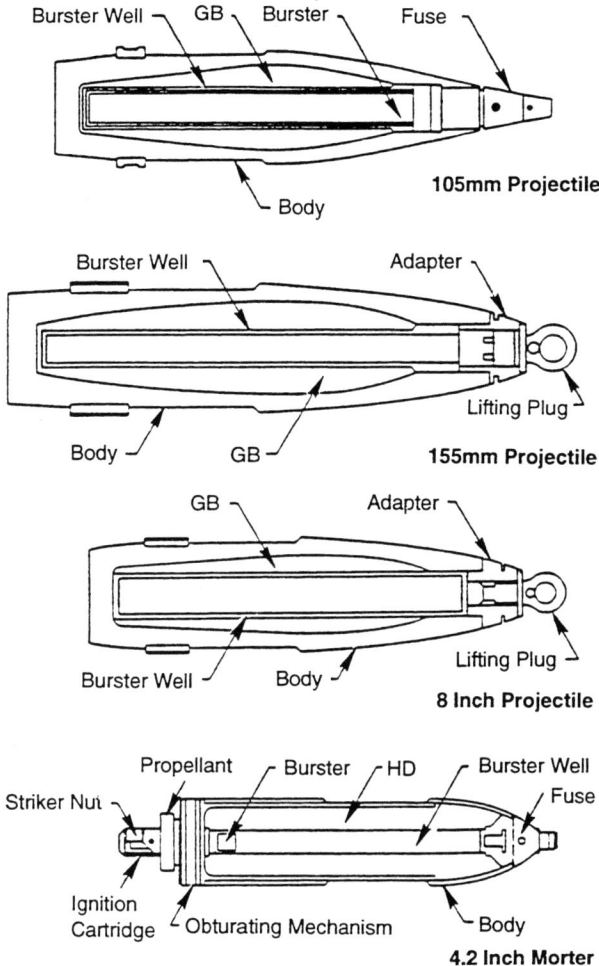

FIGURE B-2 105-mm, 155-mm, 8-inch, and 4.2-inch projectiles. Source: USATHAMA, 1982; NRC, 1993c, 1994a,c.

Geographical Distribution

The unitary chemical stockpile is located at eight continental U.S. storage sites (see figure B-4) and at Johnston Atoll in the Pacific Ocean about 700 miles southwest of Hawaii. The nature of the stockpile at each continental U.S. site, by type of container or munition and by type of agent, is indicated in table B-2.

The amount of agent, energetics, and metals stored at each site varies (table B-3). Within the continental United States, the largest quantity of chemical agent and munitions is at Tooele Army Depot, Utah, with 42.3 percent of the stockpile. All three types of agent and all types of munitions are stored there.

THE BASELINE INCINERATION SYSTEM

In this section the baseline system is briefly described. The first-generation system, JACADS, is now operating on Johnston Island, having successfully completed Operational Verification Testing (OVT) in March 1993. Figure B-5 shows the major components of the baseline system.

Storage, Transportation, and Unloading of Munitions and Containers

Munitions are stored in vented igloos, and the igloo area is monitored for agent. Most bulk containers are stored in the open or in monitored warehouses. Prior to

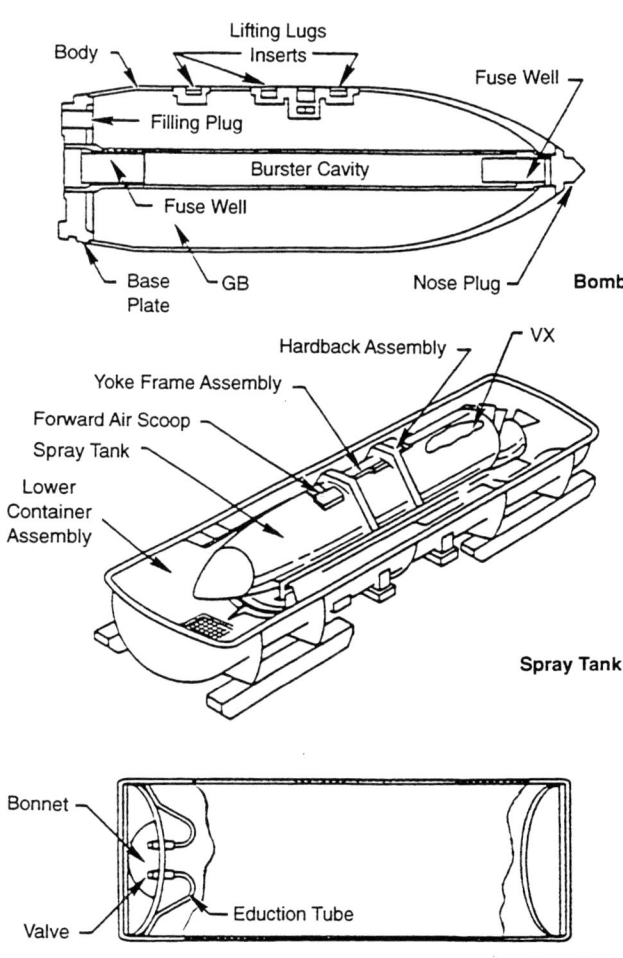

FIGURE B-3 Bomb, spray tank, and ton container. Source: USATHAMA, 1982; NRC, 1993c, 1994a,c.

TABLE B-1 Composition of Munitions in the U.S. Chemical Stockpile

Munition Type	Agent	Fuse	Burster	Propellant	Dunnage
M55 115-mm rockets[a]	GB, VX	Yes	Yes	Yes	Yes
M23 land mines	VX	Yes[b]	Yes	No	Yes
4.2-in. mortars	Mustard	Yes	Yes	Yes	Yes
105-mm cartridges	GB, mustard	Yes	Yes	Yes	Yes
105-mm projectiles	GB, mustard	Yes[c]	Yes[c]	No	Yes
155-mm projectiles	GB, VX, mustard	No	Yes[c]	No	Yes
8-in. projectiles	GB, VX	No	Yes[c]	No	Yes
Bombs (500-750 lb)	GB	No	No	No	Yes
Weteye bombs	GB	No	No	No	No
Spray tanks	VX	No	No	No	No
Ton containers	GB, VX, GA,[d] mustard, Lewisite[e]	No	No	No	No

[a] M55 rockets are processed in individual fiberglass shipping containers.
[b] Fuses and land mines are stored together but not assembled.
[c] Some projectiles have not been put into explosive configuration.
[d] GA (Tabun), or ethyl-N,N-dimethylphosphoramidocyanidate, is a nerve agent.
[e] Lewisite, or Dichloro(2-chlorovinyl) arsine, is a volatile arsenic-based blister agent.

Source: U.S. Army, 1988.

transporting munitions and containers, the area is checked for signs of leakage. If agent contamination is found, special procedures are followed to isolate and contain leaking munitions and to decontaminate the area. The munitions or ton containers are then loaded into robust, vapor-tight transport containers designed to withstand impacts and exposure to fire. (A transport container for spray tanks is yet to be designed.) The transport container is moved from the storage area to the unpacking area within the disposal building, where munitions and agent containers are unpacked manually. Packing materials (dunnage) are transported to the dunnage furnace.

Disassembly and Draining

Munitions are moved into an explosive containment room that is maintained below atmospheric pressure to prevent leakage of agent outside the enclosure and is designed to withstand overpressures that might result from the explosion of munitions during processing. Ventilation air from this room is passed sequentially through six charcoal filter beds, with agent monitors after the first, second, and fourth beds. Agent traces were rarely found after the first bed and were never detected beyond the second bed throughout the OVT at JACADS. After OVT and years of operation, some trace agent leakage through maintenance door gaskets on the carbon filtration system was detected at JACADS. Testing of improved gasket materials is under way at JACADS, and the new materials will be installed at the TOCDF prior to the start of agent operations.

Bulk storage containers are taken to a bulk drain station where they are mechanically punched and drained within an enclosure; the air of the enclosure also passes through the charcoal bed filter banks.

Agent is removed from munitions and containers by automated machinery by one of two processes. Where possible, agent storage compartment walls in M55 rockets, land mines, bombs, spray tanks, and ton containers are simply punched and drained of agent. Heavy-walled steel artillery projectiles must be disassembled. Disassembly begins with the removal of explosive elements in the case

APPENDIX B 89

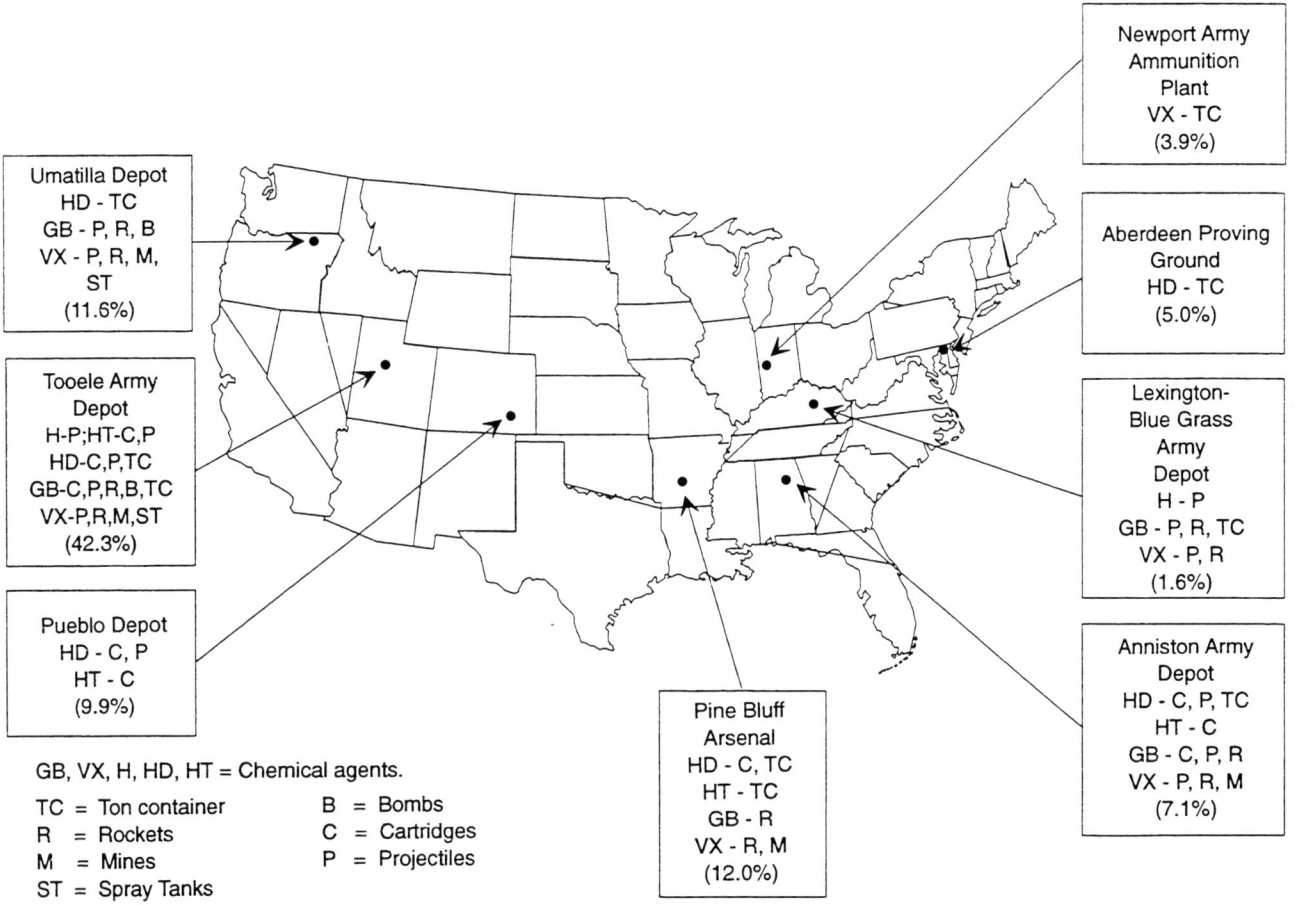

FIGURE B-4 Types of agent and munitions and percentage of total agent stockpile (by weight of agent) at each storage site. Source: OTA, 1992; NRC, 1994a,c.

of armed projectiles. In all cases, mechanical extraction of a press-fit burster well gains access to the agent. Agent drainage (and subsequent destruction) can be complicated because of gelling or solidification of the material, which then does not drain from the munition or ton container. Gelling occurs mostly in aging mustard.

These operations result in three separate streams of material that are fed to specially designed destruction systems: an agent stream that is stored in a feed tank prior to injection into the liquid incinerator; a mixed stream of energetics, small metal components, and residual agent that is fed to the rotary kiln deactivation furnace system; and large metal parts (e.g., ton containers, spray tanks, artillery projectiles), with residual agent but no energetics, that are fed to the metal parts furnace. The separation of these three streams is an important safety feature of the baseline system, enabling the designer to tailor each disposal system for specific material streams to ensure safe, controllable operations. As a result, most agent is treated in liquid form; energetics and metal parts where only residual agent is present are treated separately.

Agent Destruction

Because of the risk of earthquakes, the volume of agent stored for processing at the TOCDF has been greatly reduced (by a factor of about 5 compared to JACADS). The drained agent at the TOCDF will be stored in a 500-gallon tank inside a room designed to contain toxic substances. This tank represents the largest volume of agent in a single container on-site. A larger emergency dump tank is also provided at Tooele but is not intended to be used for normal operations.

The liquid incinerator consists of two sequential combustion chambers and a pollution abatement system

TABLE B-2 Chemical Munitions Stored in the Continental United States

Chemical Munitions (Agent)	APG	ANAD	LBAD	NAAP	PBA	PUDA	TEAD[a]	UMDA
Mustard agent (H, HD, or HT)								
105-mm projectile (HD)		X			X			
155-mm projectile (H, HD)		X	X		X	X		
4.2-in. mortar (HD, HT)		X			X		X	
Ton container (HD)	X	X			X	X[b]	X	X
Ton container (HT)					X			
Agent GB								
105-mm projectile		X				X		
155-mm projectile		X					X	X
8-in. projectile		X	X				X	X
M55 rocket		X	X		X		X	X
500-lb bomb								X
750-lb bomb							X	X
Weteye bomb							X	
Ton container		X[b]	X[b]		X[b]		X	X
Agent VX								
155-mm projectile		X	X				X	X
8-in. projectile							X	X
M55 rocket		X	X		X		X	X
M23 land mine		X			X		X	X
Spray tank							X	X
Ton container				X				

[a]Small quantities of Lewisite and tabun (GA) are stored in ton containers at TEAD.
[b]Small quantities of agent drained as part of the Drill and Transfer System assessment for the M55 rockets.

Note: APG, Aberdeen Proving Ground, Md.; ANAD, Anniston Army Depot, Ala.; BAD, Blue Grass Army Depot, Ky.; NAAP, Newport Annex Army Depot, Ind.; PBA, Pine Bluff Arsenal, Ark.; PUDA, Pueblo Depot Activity, Colo.; TEAD, Tooele Army Depot, Utah; and UMDA, Umatilla Depot Activity, Ore.

Source: Information supplied by the Program Manager for Chemical Demilitarization at a meeting of the Committee on Alternative Chemical Demilitarization Technologies, March 9–10, 1992, National Academy of Sciences.

(discussed below). The first, or "primary," combustion chamber is preheated to an operating temperature of 2,700°F with fuel before agent is injected. The primary fuel is natural gas; liquified propane gas is stored in an on-site tank to provide a backup fuel supply. As agent flow increases, the fuel flow is decreased to maintain the desired temperature for effective agent destruction. Agent flow to the burner is stopped if the temperature drops below 2,550°F. Gases from the first chamber are sent to a secondary chamber, also preheated with fuel, for a final burn stage at 2,000°F. The afterburner gases are then treated in the pollution abatement system.

Some slag produced during nerve agent destruction will form on the lower-temperature walls of the secondary chamber. Spent decontamination fluid is also injected into the secondary chamber to ensure destruction of any residual agent in the solution as well as the evaporation and discharge of the water vapor. This fluid also contains salts that are deposited in the bottom of the secondary chamber. The liquid incinerator at JACADS had to be shut down periodically for manual removal of glasslike solidified salts from both agent and decontamination fluid disposal. A slag removal system has been developed to discharge molten salts during operations at the TOCDF.

Destruction of Energetics

Energetics (fuses, boosters, bursters, and solid rocket propellant) are burned in a counterflow rotary kiln (deactivation furnace system). Energetics are all contained

TABLE B-3 Approximate Amounts of Metals, Energetics, and Agent Contained in the Unitary Chemical Stockpile (tons), by Site

Site	Ferrous Metal	Aluminum	Explosive	Propellant	Estimated Agent[a]
Tooele	22,000	570	350	175	10,500
Anniston	13,700	1,020	451	757	1,800
Umatilla	7,930	1,380	338	1,030	2,900
Pine Bluff	2,644	1,431	180	1,060	3,000
Lexington	1,631	904	115	670	400
Pueblo	10,910	0	124	0	2,500
Newport	2,455	0	0	0	1,000
Aberdeen	NA[b]	0	0	0	1,300
JACADS	NA	NA	NA	NA	1,700
TOTAL	61,270	5,305	1,558	3,692	24,800

[a]Estimated values, calculated by the Alternatives Committee, based on percentages of the total stockpile at each site, multiplied by 25,000 tons.
[b]NA—not available.

Source: Information supplied by the Program Manager for Chemical Demilitarization at a meeting of the Committee on Alternative Chemical Demilitarization Technologies, March 9–10, 1992, National Academy of Sciences.

in thin-walled metallic housings that must be punched or cut into pieces prior to burning; confined energetics would detonate in the kiln rather than burn. M55 rockets, after being drained of agent, are sliced into eight pieces to expose energetic material surface area so the material will burn without detonating. Draining and slicing are both done while the rocket is in its fiberglass launch tube. Bursters from artillery projectiles are also sliced, but after removal from the projectile. Explosive elements in land mines are punched in place to expose the explosive and are not removed from the munition. The pieces, most of which may be wetted with agent, are fed slowly into the downstream end of the kiln (downstream in the sense of gas flow) to avoid explosive concentrations within the kiln. Solid pieces move upstream (against the gas flow) as the energetics are burned and then moved onto an electrically heated discharge conveyor, where the temperature is maintained at 1,000°F for 15 minutes. This results in a "5X" decontaminated material, which is the Army's classification for material that is suitable for release to the public.

The resultant mixture of light steel components, melted aluminum, and glass fibers is of no commercial value. Gases discharged from the rotary kiln pass through an afterburner where they are subjected to a temperature of 2,200°F for 2 seconds. This is a higher temperature and longer time than was used for oxidation at JACADS (2,000°F for 1 second) and should ensure that the TOCDF furnace fully complies with requirements for the complete destruction of polychlorinated biphenyls (PCBs), small quantities of which are present in some fiberglass launch tubes. The afterburner gases are then treated in the pollution abatement system.

Metal Parts Decontamination

Metal parts that have been drained of agent (ton containers, bombs, spray tanks, artillery projectiles, and burster wells, which were pulled to access the agent) are heated to 1,000°F and maintained at that temperature for 15 minutes in a fuel-fired metal parts furnace to produce metal suitable for release as scrap (defined by the Army as 5X). Residual or undrained (including gelled) agent that has not been removed is vaporized and burned within the furnace. This process takes additional time and can limit the system's throughput. At JACADS,

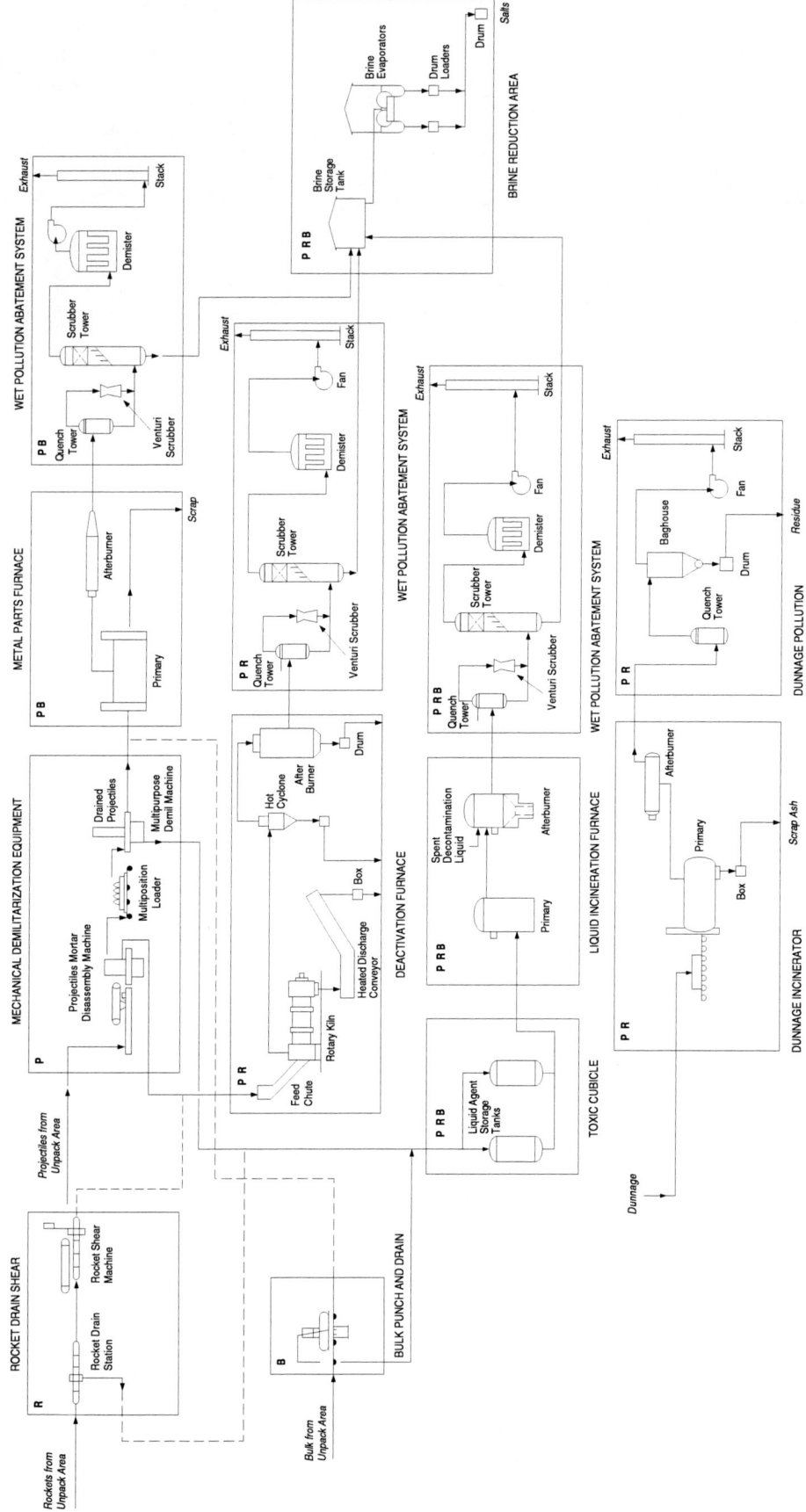

FIGURE B-5 Schematic drawing of the baseline system. Source: U.S. Army, 1988; NRC, 1994a,c.

special procedures were approved by the Environmental Protection Agency and implemented to increase the quantity (over the design limit of 5 percent residual per ton container) of agent processing in the metal parts furnace. This procedure ensured compliance with the specifications in the Resource Conservation and Recovery Act (RCRA) permit (U.S. Code of Federal Regulations, 1976). After testing, this modification has been shown to be acceptable, with proper monitoring and control, but the RCRA permit should be clarified so that waivers will not be required for operation at the TOCDF. Gases discharged from the metal parts furnace are passed through an afterburner, maintained at 2,000°F, before being treated in the pollution abatement system.

Pollution Abatement Systems

The liquid incinerator, deactivation furnace system, and metal parts furnace employ identical, dedicated pollution abatement systems. Gases leaving the secondary chamber of the liquid incinerator or the metal parts furnace afterburner flow to separate dedicated pollution abatement systems for removal of gaseous pollutants and particles to meet emission standards. Hot gases leaving the deactivation furnace system kiln flow to a refractory-lined cyclone separator, where large particles (glass fibers from rocket launch tubes) are removed; next, the gases enter the afterburner; finally, they flow to a similar pollution abatement system.

Each pollution abatement system consists of a quench tower, a venturi scrubber, a packed bed scrubber, a candle mist-eliminator vessel, brine or quench recycle pumps, and an induced draft (ID) blower. Figure B-6 is a schematic drawing of a pollution abatement system.

The exhaust gas stream enters the quench tower near the bottom, where it is cooled by contact with a countercurrent spray of brine pumped from the packed bed scrubber sump. Acidic or acid-forming gases (e.g., hydrogen chloride, hydrogen fluoride, nitrogen oxides (NO_x), carbon dioxide, and sulfur dioxide, depending on the chemical agent incinerated)

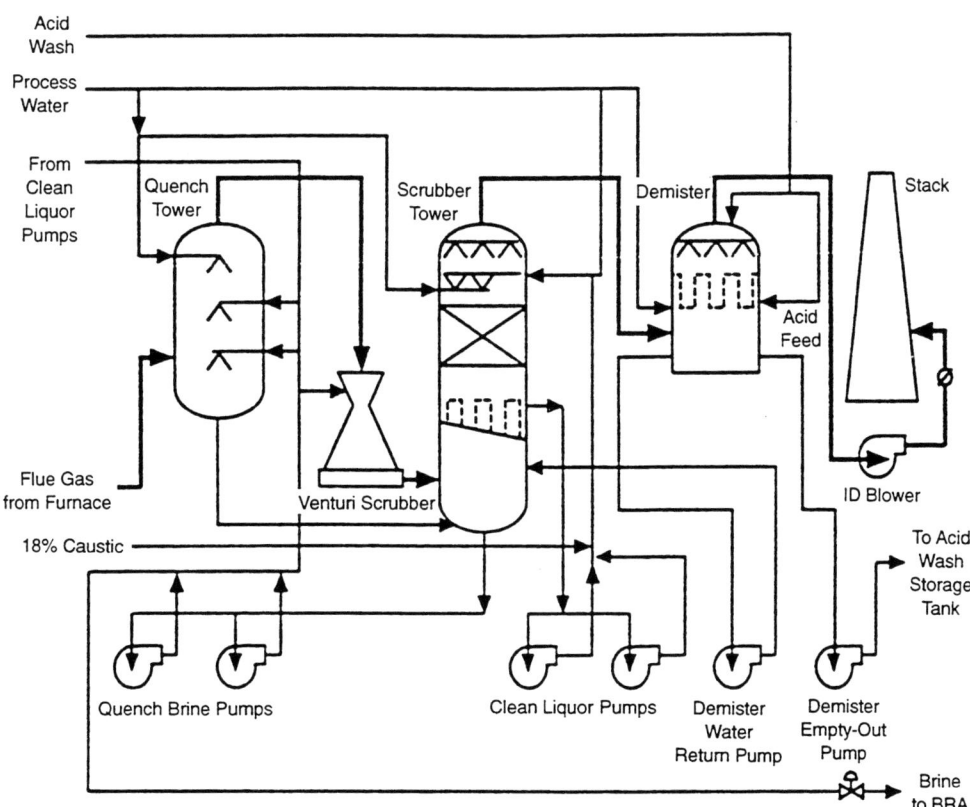

FIGURE B-6 Schematic drawing of a pollution abatement system. Source: MITRE, 1993a; NRC, 1994a,c.

in the exhaust gas react with the caustic brine to form salts, which remain in solution in the brine. The cooled gas stream exits from the top of the quench tower and enters a variable throat venturi where it is scrubbed to remove particulates. The venturi has a variable throat to maintain a constant pressure drop independent of the flow of exhaust gases. The brine streams from the quench and venturi scrubber are then returned to the scrubber tower sump. Process water is added to the packed bed scrubber sump to make up for water evaporated in the quench tower. An 18 percent caustic (sodium hydroxide) solution is added, as necessary, to the sump to maintain the brine pH above 8 or 9.

The exhaust gas stream from the venturi scrubber enters the scrubber tower below the clear liquor reservoir tray, moves upward through the packed bed section, and exits at the top of the tower, after passing through a mist-eliminator pad. In the packed bed section, the gas stream comes in contact with a brine solution flowing countercurrently through the bed. Acidic gases in the exhaust gas stream are further scrubbed with caustic brine. The brine solution from the packed bed falls back to the reservoir tray and is recycled back to the top of the packed bed section. Excess brine overflows into the tower sump. Brine density is controlled by pumping a brine stream into the brine reduction area (BRA) storage tanks and replacing it with processing water.

The scrubbed gases enter a candle mist-eliminator vessel. Mist-eliminator candles (i.e., candle-shaped fabric filters) remove very fine mist and submicron particulate matter that were not removed in the venturi scrubber. The cooled and cleaned exhaust gases are pulled through an induced draft blower located upstream of the stack shared by the three pollution abatement systems.

In the tests at JACADS, particulate emissions from the liquid incinerator, the deactivation furnace system, and the metal parts furnace were consistently low (the dunnage furnace was not tested). The mean particulate concentration for all trial burns for each incinerator was less than 5 mg/m^3 at 7 percent oxygen, with a maximum value of 10.9 mg/m^3. Permits require less than 180 mg/m^3. The tests show that metal emissions are extremely low, frequently below detectable limits.

Auxiliary Systems

The dunnage furnace and its pollution abatement system consist of a feed handling system, a primary chamber, an afterburner, a quench tower, a bag house separator, an induced draft blower, and a separate exhaust stack. This system is designed to burn both noncontaminated and contaminated dunnage from the

TABLE B-4 Air and Exposure Standards

	Permissible Hazard Levels in Air (mg/m^3)			Lethal Human Doses		
Agent	Workers[a]	Stack Emissions[b]	General Population[c]	Skin, LD_{50} (mg/kg)	Intravenous, LD_{50} (mg/kg)	Inhalation, LCt_{50} (mg-min/m^3)
GA	0.0001	0.0003	0.000003	14-21	0.014	135-400
GB	0.0001	0.0003	0.000003	24	0.014	70-100
VX	0.00001	0.0003	0.000003	0.04	0.008	20-50
H/HD/HT	0.003	0.03	0.0001	100		10,000

Note: The Army standards shown in the first three columns set the minimum level of performance required for gas release by any alternative process and are applicable to all four process streams. LCt_{50} and LD_{50} represent dosage and dose, respectively, that result in 50 percent lethality. LCt_{50} represents a concentration (mg/m^3) times the exposure time (min).

[a] For 8-hour exposure.
[b] Maximum concentration in exhaust stack.
[c] For 72-hour exposure.

Source: U.S. Army, 1974, 1975, 1988; NRC, 1993c.

munitions processing operations, as well as charcoal and high-efficiency particulate air (HEPA) filter media from the air filters. Exhaust gases from the afterburner flow into the dunnage pollution abatement system quench tower. A water quench is used to cool the exhaust gases, and a bag house is used to remove particles. This pollution abatement system does not include acid gas scrubbing. The exhaust gases are maintained above the saturation temperature to prevent moisture from condensing in downstream equipment. Gases exhaust to the atmosphere through a separate stack via the dunnage furnace induced draft blower.

Initially, problems demonstrating acceptable performance of this unit at JACADS prevented incineration of most of the dunnage there. The alternative at JACADS has been to dispose of materials as hazardous waste. If the dunnage incinerator at Tooele is not proven satisfactory, an alternative dunnage waste disposal strategy must be developed and proven prior to agent operations there. The Army has decided not to burn demilitarization protective ensemble (DPE) suits (containing polyvinyl chloride) from Tooele operations in the dunnage furnace to avoid public concerns about the potential of chlorinated dioxins and furans in the exhaust.

In the brine reduction area, discarded process brines are collected, stored, and evaporated, and salts from the pollution abatement systems for the three furnaces are dried. Operation of the brine reduction system produces salt that contains 10 percent or less water by weight. The brine reduction system consists of four subsystems: (1) steam generation (boilers); (2) brine evaporation; (3) brine drying; and (4) pollution abatement. Entrained particles from the brines are collected in a bag house before exhaust is discharged to the atmosphere. The brine reduction area pollution abatement system consists of a heated, dual-module bag house dust collection system. A fan pulls the exhaust gas through the bag house modules prior to discharge to the atmosphere through a stack. Brine reduction area exhaust is heated in a fuel-fired superheater so that the exhaust remains above the dew point as it passes through the filters in the bag house modules. The bag house modules are equipped with a pulse air jet system that cleans the bags continuously. As solids accumulate in drums under the bag house, they are packaged and stored for shipping to land disposal sites as hazardous waste.

Satisfactory operation of the brine reduction area was not demonstrated during the OVT. Modifications to this system are described in chapter 2.

Agent Monitoring Systems

The agent monitoring systems to be installed at Tooele are the same as the systems at JACADS. There are two types of analyzers: (1) the Automatic Continuous Air Monitoring System (ACAMS), which is capable of detecting agent at concentrations well below the levels that present an immediate threat to plant personnel or the surrounding population, with a response time of three to eight minutes; and (2) the Depot Area Air Monitoring System (DAAMS) for collecting longer, time-averaged samples for more selective subsequent analysis in the laboratory. The ACAMS monitors in personnel areas and in the stack are set to trigger an alarm at 20 percent of permissible agent levels (table B-4), at which point agent operations are shut down. The DAAMS samples are analyzed for the much lower permissible general population levels. ACAMS and DAAMS monitoring points are distributed throughout the facility at appropriate locations.

In the event of agent release, the ACAMS monitors provide alarms and initiate corrective actions. For example, if agent is detected in a furnace effluent, agent feed to that furnace is stopped automatically. The DAAMS system serves the dual purpose of providing samples to confirm or refute ACAMS alarms (which are sometimes false) and of documenting concentrations of agent at much lower levels of detection sensitivity. Both systems use the principle of drawing gas through a gas chromatograph equipped with a flame photometric detector. Every detection of agent is interpreted by computer analysis.

The monitoring systems must be readjusted for each agent type. ACAMS monitors generate frequent false alarms because they cannot adequately differentiate agent from other commonly encountered organic contaminants (e.g., fuel contaminants, diesel exhaust, antifreeze). For example, during 151 days of testing in the fourth set of operational verification tests, there were 55 alarms suggesting that allowable stack concentrations had been exceeded. All 55 were determined to be false positives. The retrieval and laboratory analysis of the DAAMS collection tubes to verify conditions typically require at least 30 minutes. Frequent false alarms pose several problems. They may make operators complacent and reluctant to stop operations, particularly when faced with production goals. At continental U.S. sites, false alarms could erode public confidence in the safety of the facility.

Appendix C

Recommendations of the Committee on Review and Evaluation of the Army Chemical Stockpile Disposal Program (Stockpile Committee)

Appendix C consists of four tables presenting extract listings of recommendations from 1993 and 1994 reports (NRC, 1993b; 1994a,b; 1993a; 1994c respectively) prepared by the Committee on Review and Evaluation of the Army Chemical Stockpile Disposal Program (Stockpile Committee). The fourth table (table C-4) includes findings as well as recommendations. An alpha-numeric code reference has been added to each finding and recommendation to assist the reader. These code references are applied to each use in the text.

TABLE C-1 Recommendations from *Evaluation of the Johnston Atoll Chemical Agent Disposal System Operational Verification Testing: Part I* (OVT1) and *Part II* (OVT 2)

Alpha-numeric Code	Recommendation
OVT1-1	The Army should initiate systemization of the Tooele Chemical Disposal Facility at Tooele Army Depot, Utah.
OVT1-2	The Army should use systemization of the Tooele Chemical Disposal Facility to implement improvements relating to safety, environmental performance, and plant efficiency. These improvements should be made at Tooele prior to initiating the destruction of agent and munitions.
OVT2-1	Give safety considerations priority over production goals.
OVT2-2	Proceed with Tooele systemization, and during systemization, conduct needed testing and improvement activities, including the following:
OVT2-2A	Develop and demonstrate an improved agent monitoring and identification system.
OVT2-2B	Complete the brine reduction area (to include its pollution abatement system) performance tests, or develop a satisfactory brine disposal alternative.
OVT2-2C	Demonstrate the dunnage furnace performance with various levels of chlorinated waste; if needed, either modify the pollution abatement system design (e.g., add acid gas scrubbing) or limit feed materials to those that can be handled by the existing design; alternatively, satisfactory land disposal options must be identified.
OVT2-2D	Review the probable levels of NO_x production from VX destruction and the allowable emission levels at the other continental U.S. sites requiring VX destruction; if appropriate, develop needed NO_x abatement systems.
OVT2-2E	Develop and demonstrate the proposed hot-slag removal system for the liquid incinerator system.

APPENDIX C

Alpha-numeric Code	Recommendation
OVT2-2F	Eliminate furnace feed errors by improved monitoring and control of the deactivation furnace and metal parts furnace feed systems and by improved methods for tracking the various types of munitions.
OVT2-2G	Address all problems associated with residual gelled mustard, in particular, the use of suited personnel to perform functions that were intended to be automated.
OVT2-3	Establish and maintain close working relationships with permitting agencies, and support these efforts with careful analysis of operating parameters to ensure that permits provide for safe destruction of agent, adherence to regulatory requirements, and effective plant operations.
OVT2-4	Establish programs, procedures, and management oversight to ensure continuing compliance with all environmental regulations.
OVT2-5	Develop systems to improve overall management of safety.
OVT2-6	Complete the risk assessment for the Tooele Chemical Agent Disposal Facility during the systemization period.

Source: NRC, 1993b; NRC, 1994a.

TABLE C-2 Recommendations from *Review of Monitoring Activities Within the Army Chemical Stockpile Disposal Program* (MON)

Alpha-numeric Code	Recommendation
	General recommendations
MON-1	The Army should initiate a substantial program to upgrade the monitoring systems for continental U.S. sites.
MON-2	The Army should obtain expert help at both the systems design and the equipment selection levels, perhaps by engaging a contractor with extensive experience in monitoring of trace species and in advanced instrument development.
MON-3	The Army should undertake whatever instrument development is necessary to ensure that improved instrumentation is available to the chemical disposal program in suitably rugged and operational forms.
MON-4	The Army should test and use new monitoring instrumentation at JACADS before such instrumentation is employed at Tooele.
MON-5	The Army should plan to continually improve the monitoring system in areas where performance is presently limited by unavailability of suitable instrumentation.
	Recommendations for agent/nonagent monitoring
MON-6	Add the capability for positive identification of chemical agent species (chemical speciation) to the agent detection system and analytical laboratories at all of the disposal facilities in order to reduce the occurrence of false positives.
MON-7	Institute continuous monitoring for all agents present at each facility, including those in storage areas.
MON-8	Reduce the time required for confirmation of false positives.
MON-9	Evaluate the procedures for periodic testing of field sensors to ensure that false negatives are not possible if a significant release should occur.
MON-10	Implement monitoring designed to provide more rapid response to high-level agent release.
MON-11	Evaluate the benefits of more frequent analysis of facility stack gases for nonagent trace contaminants.
	Recommendations for laboratory operations
MON-12	Increase the automation of sample handling and laboratory operations to ensure better quality control and efficiency.
MON-13	Give laboratory personnel a variety of tasks that ensure optimal attention and performance.
MON-14	Give blind challenges to the laboratory.
MON-15	Perform a detailed error analysis of the laboratory system and procedures.

Source: NRC, 1994b.

TABLE C-3 Recommendations from the letter report to the Assistant Secretary of the Army to recommend specific actions to further enhance the CSDP risk management process (RISK)

Alpha-numeric Code	Recommendation
RISK-1	A site-specific, full-scope, scenario-based risk assessment should be performed for each continental U.S. facility, starting with the Tooele facility.
RISK-2	Each site-specific risk assessment should include the case of continued storage without disposal as one scenario.
RISK-3	The risk assessments should be quantitative and include the following features: • bottom-line results on the health effects to on-site personnel in terms of likelihood and consequence, including a site-specific atmospheric dispersion and health effects analysis and an analysis of emergency response capability; • a clearly defined set of scenarios that, taken together, provide a comprehensive representation of the risk; • dependency matrices that display inter- and intra-system dependencies; • a human action analysis that represents the human role in controlling risk; • quantification of risk from all causes, including both internal events (plant and plant-people failures) and external events (earthquakes, fires, floods, aircraft crashes, etc.); • site-specific hardware, software, procedures, training programs, maintenance practices, and operations personnel (including site-specific storage facilities and munitions handling activities); • risk contributors in such terms as random failures, common cause failures, multiple failures, and human error; and • an uncertainty analysis to display clearly how much confidence the analysts have in the precision of the quantitative results.
RISK-4	Modern, up-to-date methodologies should be employed, such as those found in the risk assessments reported in NUREG-1150.
RISK-5	The risk assessments should be conducted by organizations with recognized expertise in the field, but not otherwise involved in the CSDP. In a similar vein, independent peer reviews are an absolute requirement.
RISK-6	Local representatives of neighboring communities must be involved early. Their concerns about the CSDP may be substantial, and will warrant consideration throughout the analysis process.
RISK-7	Emphasis must be placed on human reliability factors, particularly in light of the human factors issues raised by the Stockpile Committee in reviewing the first phase of Operational Verification Testing at JACADS.
RISK-8	To avoid overstatement of the results it is important that the confidence levels of the risk parameters be fully displayed. It is this process of quantifying the uncertainty in the risk that will establish the reliability of the conclusions. Experience has indicated that the results of a risk assessment provide valuable information on the importance of different contributors to risk, not only in terms of hardware failures but also in terms of human errors and deficiencies in procedures and software. Thus the risk assessment can lead to process changes that reduce overall risk.

Source: NRC, 1993a.

TABLE C-4 Recommendations (REC) and Findings (FIND) from *Recommendations for the Disposal of Chemical Agents and Munitions*

Alpha-numeric Code	Finding and Recommendation
	Expeditious Progress
FIND/REC-1	The storage risk will persist until disposal of all stockpile materials is complete. Both storage risk and disposal risk will increase with time as the stockpile deteriorates further. Existing analyses indicate that the annual storage risk to the public at each site is the same as or greater than the annual risk due to disposal. Thus, total risk to the public will be reduced by prompt disposal of the stockpile.
REC-1	The Chemical Stockpile Disposal Program should proceed expeditiously and with technology that will minimize total risk to the public at each site.
	Risk Analyses
FIND/REC-2	Existing risk analyses did not evaluate the latent health hazards associated with storage, handling, and disposal activities. These latent risks represent one of the major concerns voiced by the public.
REC-2	The committee expects the latent risks from storage, handling, and disposal activities to be low. However, new risk analyses should be conducted that explicitly account for latent health risks from storage, handling, and disposal.
FIND/REC-3	The finding that total risk will be reduced by prompt disposal, although apparently reasonable, is based upon earlier analyses that do not reflect current risk assessment methods and knowledge about the storage, handling, and disposal activities.
REC-3	Updated analyses of the relative risk of storage, handling, and disposal activities should be completed as soon as possible.
FIND/REC-4	The Stockpile Committee is confident that site-specific risk analyses will confirm the wisdom of proceeding promptly. Further, the schedule of the disposal program should not be delayed pending completion of the updated analyses, because they can be conducted concurrently with other activities within the overall construction and operations schedule. Both storage risk and processing risk differ from site to site. Storage risks differ greatly depending on storage configuration, types and mix of munitions, and the potential for external events, as well as nearby community conditions.
REC-4A	New risk analyses should be site specific, using the latest available information and methods of analysis. At this time, since there is insufficient knowledge of potential alternative technologies, a first-cut series of analyses should compare the relative risks of continued storage and disposal by the baseline system. Analyses should identify the major contributors to total risk including storage. The analyses will confirm or refute the present belief that maximum safety dictates prompt disposal.
REC-4B	As new, site-specific risk analyses become available, the Army should reconsider the schedule of construction and operation of disposal facilities and, if indicated, reorder the remaining sequence so as to minimize any subsequent cumulative total risk. The Army should also consider reconfiguring each high-risk stockpile to a safer condition prior to disposal if this will significantly decrease cumulative total risk.

TABLE C-4 Recommendations (REC) and Findings (FIND) from *Recommendations for the Disposal of Chemical Agents and Munitions*

Alpha-numeric Code	Finding and Recommendation
FIND/REC-5	The committee does not foresee that any alternative agent destruction technology will substantially reduce the total agent processing risk. Site-specific risk analyses will identify the potential to improve safety over the baseline system and thus serve as a check on this belief.
REC-5	As research progresses on potential alternative technologies and as their potential for improved safety becomes apparent, site-specific risk analyses should be reexamined, with the potential alternative substituted in the baseline system, to estimate overall system performance. In view of the limited potential for overall safety improvement, however, the disposal program should not be delayed pending completion of such research.

Public Concerns

FIND/REC-6	The members of the public in communities near the chemical stockpile sites have voiced diverse views and opinions regarding the stockpile disposal program, and their desire to have greater access and input into decisions concerning that program. The committee's public forum, as well as correspondence and telephone calls to the committee, indicate that the Army is not as well informed of public sentiment as desirable. The public wants a larger role in the selection of disposal technology, the monitoring of operations that ensure its own safety, and determining the fate of the facility after completion of disposal efforts.
REC-6	The Army should develop a program of increased scope aimed at improving communication with the public at the storage sites. In addition, the Army should proactively seek out greater community involvement in decisions regarding the technology selection process, oversight of operations, and plans for decommissioning facilities. Finally, the Army should work closely with the Chemical Demilitarization Citizens Advisory Commissions, which have been (or will be) established in affected states. There must be a firmer and more visible commitment to engaging the public and addressing its concerns in the program.

Current System

FIND/REC-7	Chemical agents and munitions materials have been successfully divided into four distinct process streams having widely differing properties. Separation of these materials for processing in distinct, well-engineered systems provides a safer and more reliable operation than would processing of a mixed stream in a single process.
REC-7	All disposal systems should be designed to separately process agent, energetics and associated small metal components, large metal parts, and dunnage streams.
FIND/REC-8	The committee found no acceptable alternative to mechanical methods to gain access to agent in munitions and to separate agent, energetics, and associated small metal components, and large metal parts.
REC-8	The Army should continue with mechanical methods to gain access to agent and to separate material streams. Alternative mechanical systems should be pursued if simpler, more durable concepts, which also permit separation of the streams, are discovered.
FIND/REC-9	Gelled agent, particularly mustard, is difficult to separate from its container and will hamper any agent destruction or neutralization process or any attempt to decontaminate containers.
REC-9	Research to develop means to extract, handle, and process gelled agents should be accelerated, to sustain the advantages of handling separate streams and to facilitate the use of alternative technologies.

TABLE C-4 Recommendations (REC) and Findings (FIND) from *Recommendations for the Disposal of Chemical Agents and Munitions*

Alpha-numeric Code	Finding and Recommendation
FIND/REC-10	The committee found no readily applicable alternative to incineration of energetic components. Energetics are solid materials, cast in place in metal containers. In this form they are not compatible with alternative oxidation technologies that require liquid or finely divided feed materials. Extraction of energetics and reduction to suitable slurry form would be difficult and hazardous.
REC-10	Dispose of energetic materials by incineration.
FIND/REC-11	The committee found no alternative to high-temperature treatment for reliable decontamination of metal parts to a level suitable for release to the public.
REC-11	Use of the baseline metal parts furnace or other high-temperature treatment is recommended.
FIND/REC-12	The Johnston Atoll Chemical Agent Disposal System (JACADS) Operational Verification Testing (OVT) provided additional assurance that the baseline system is capable of the safe disposal of the Army's chemical stockpile. However, the committee found that OVT identified opportunities for improvements in operations, management practices, and training with regard to safety, environmental performance, and plant efficiency. The committee has recommended that systemization be used to implement these improvements prior to the initiation of the destruction of agent and munitions at Tooele.
REC-12	The Chemical Stockpile Disposal Program should continue on schedule with implementation of the baseline system, unless and until alternatives are developed and proven to offer safer, less costly, or more rapidly implementable technologies (without sacrifice in any of these areas). Baseline system improvements should be implemented as identified and successfully demonstrated.
FIND/REC-13	The Stockpile Committee finds the baseline system to be adequate for disposal of the stockpile. Addition of activated carbon filter beds to treat all exhaust gases would add further protection against agent and trace organic emissions, even in the unlikely event of a substantial system upset. If the beds are designed with sufficient capacity to adsorb the largest amount of agent that might be released during processing, addition of these beds could provide further protection against inadvertent release of agent.
REC-13	The application of activated charcoal filter beds to the discharge from baseline system incinerators should be evaluated in detail, including estimations of the magnitude and consequences of upsets, and site-specific estimates of benefits and risks. If warranted, in terms of site-specific advantages, such equipment should be installed.
	Alternatives
FIND/REC-14	After examination of all the technologies brought to the attention of the Stockpile Committee by the Alternatives Committee and others, the Stockpile Committee has determined that four neutralization-based systems offer the most promise for agent destruction. Neutralization has been demonstrated to be effective for GB but is not yet proven for mustard and VX. Utilizing lower temperatures and pressures and ordinary chemical processing equipment, neutralization is simpler than incineration, and it may be lower in cost for some sites. Recent laboratory studies have reported encouraging results for the neutralization of neat VX and mustard (see Appendix E [of source document]), though questions remain for neutralizing impure and gelled materials. Reaction products from neutralization processes will require further treatment prior to disposal. Potentially applicable processes for further treatment of these reaction products are incineration, wet air oxidation, supercritical water oxidation, and biological treatment. All of these combinations will require further research and demonstration to ensure that the combination of these processes treats agent to levels consistent with treaty and environmental requirements. In view of the increasing total risk associated with disposal program delays, and recognizing that public opposition might delay the program for a number of reasons, including opposition to incineration, it is imperative that alternative technologies be developed promptly.

TABLE C-4 Recommendations (REC) and Findings (FIND) from *Recommendations for the Disposal of Chemical Agents and Munitions*

Alpha-numeric Code	Finding and Recommendation
REC-14A	Neutralization research should be substantially accelerated and expanded to include field-grade and gelled material as appropriate and the neutralization of drained containers.
REC-14B	Neutralization research should be accompanied by preliminary analyses of integrated systems capable of reducing agents all the way to materials acceptable for transport or disposal.
REC-14C	These analyses and research should be conducted in parallel to lead to the selection of a single system for further development.
FIND/REC-15	There has been continued development of various research programs involving potential alternatives since the National Research Council report *Alternative Technologies for the Destruction of Chemical Agents and Munitions* was issued.
REC-15	The Army should continue to monitor research developments in pertinent areas.
FIND/REC-16	Neutralization of agent and decontamination of containers, followed by transport of both to another facility for final treatment, offer an attractive alternative to the baseline liquid incinerator, especially for sites with no stored energetics. Receiving sites might be another chemical agent disposal site or commercial hazardous waste incineration facilities (if possible). This option could be viable at Newport Army Ammunition Plant and at Aberdeen Proving Ground, provided complications with gelled mustards do not arise.
REC-16	Neutralization followed by transport for final treatment should be examined as an alternative, at the Aberdeen and Newport sites. This examination should include location of acceptable receiver sites and transport routes, and a comparison of costs and schedules relative to on-site baseline treatment. If favorable results are indicated, the examination should be expanded as an option to eliminate the liquid incinerator at other sites. At those locations, on-site incineration of energetics and associated metal parts is still recommended.
FIND/REC-17	The current chemical stockpile disposal schedule may provide time for site-specific substitution or integration of proven alternative agent disposal processes at selected sites if research and development efforts are accelerated and results are favorable.
REC-17	Proven alternative technologies, if available without increasing risk, should be considered for application on the basis of site-specific assessments.
FIND/REC-18	Future developments for the baseline system as well as for a number of alternative technologies will require a flexible, agent-qualified experimental facility.
REC-18	The facility and staff at the Chemical Agent Munitions Disposal System (CAMDS) facility should be maintained at an effective operating level for the foreseeable future. However, agent stocks should not be deliberately retained at Tooele in order to feed an alternative technology demonstration.
FIND/REC-19	Application of all known alternative agent disposal systems will require research and development, and demonstrated safe operation (operational verification testing) with chemical agents.
REC-19	Application of an alternative technology at any site should be preceded by demonstration of safe, pilot operation (operational verification testing) at the Chemical Agent Munitions Disposal System facility. These operations should not be carried out on a trial basis at storage sites.

TABLE C-4 Recommendations (REC) and Findings (FIND) from *Recommendations for the Disposal of Chemical Agents and Munitions*

Alpha-numeric Code	Finding and Recommendation
	Stockpile Safety
FIND/REC-20	A recent MITRE Corporation evaluation of stockpile condition with respect to propellant stabilization in M55 rockets suggests that the stockpile is safe until 2007 or later, whereas a similar Army report suggests 2002. The MITRE report notes that stockpile surveillance may be reduced in the belief that the stockpile will be disposed of by 2004. The committee is concerned that there is considerable uncertainty in all of the attempts to estimate safe storage life of the M55 rocket propellant. Degradation is not well understood. If surveillance is reduced, it would leave the stockpile subject to dangerous uncertainty. Further, other signs of degradation—gelled mustard, foaming mustard artillery shells, leaking and corroded ton containers—suggest that stockpile degradation can adversely affect disposal processes. Finally, realistic estimates of the duration of the disposal effort will extend well beyond 2004, particularly if alternative technologies are to be used.
REC-20	Further research into the nature and sequence of propellant stabilizer degradation should be undertaken promptly. The present condition of the stockpile should be evaluated with sufficient new field sampling of propellant grains, including grains from representative leakers that have been overpacked. Stockpile surveillance should be increased rather than decreased, particularly for M55 rockets.
	Staffing Needs
FIND/REC-21	The Army faces significant challenges in executing the Chemical Stockpile Disposal Program. As more sites begin development, important engineering and technical issues will be faced. These will cover a large spectrum over the life of this program, and will include, for example, development and maturation of alternative technologies, as well as development of a method for extracting and disposing of gelled mustard. These challenges will create more demand for planning, management, and supervision than the office of the Program Manager for Chemical Demilitarization will be capable of providing without augmentation. A shortage of skilled staff could have safety implications for the program, as well as its more obvious implications for program slowdown with attendant increased risk.
REC-21	The Army should establish a program to incrementally hire (or assign military) personnel to ensure that staff growth is consistent with the workload and with technical and operational challenges. These additional personnel must be assigned and trained before the project office gets deeply involved in addressing each challenge.

Source: NRC, 1994c.

Appendix D

Public Meeting
Tooele County Courthouse, Tooele, Utah

On March 29, 1995, the National Research Council Committee on Review and Evaluation of the Army Chemical Stockpile Disposal Program (Stockpile Committee) held a public meeting at the Tooele County Courthouse in Tooele, Utah, for the purpose of receiving briefings from and holding discussions with the Utah Chemical Demilitarization Citizens Advisory Commission (CAC), the state Division of Comprehensive Emergency Management, and the state Division of Solid and Hazardous Waste. The committee desired to learn about the perspectives of these organizations with regard to the Army's Chemical Stockpile Disposal Program. The agenda for this meeting is reproduced as part of this appendix.

Letters of invitation were sent to all three organizations. The letter to the Citizens Advisory Commission is reproduced to show the content of the letter and the list of individuals and other government agencies to which all letters were copied.

The Stockpile Committee also dispatched invitations to almost 100 other state and local officials and private individuals. These letters extended an invitation to appear personally before the committee, as well as to provide written materials. The list of individuals was drawn up with the assistance of state and local officials, as well as with the help of private citizens and organizations. There was a diligent attempt by the National Research Council staff to include all interested individuals and groups on the list. An example of these letters of invitation is also reproduced in this appendix (see Public Invitation Letter), as well as a list of all agencies and persons to whom the letters were sent (see Distribution List).

AGENDA

WEDNESDAY, MARCH 29, 1995
Tooele County Courthouse, 47 South Main Street, Tooele, Utah

CITIZENS MEETING:	**COMMUNITY INVOLVEMENT SUBGROUP**
COMMITTEE:	Dr. Ann Fisher, *Lead* Dr. Richard Magee Dr. Dennis Bley Dr. Walter May Dr. Elisabeth Drake Dr. Alvin Mushkatel Mr. Gene Dyer Mr. Peter Niemiec
NRC STAFF:	Mr. Bruce Braun, Director, BAST Mr. Archie L. Wood, Executive Director, CETS Mr. Donald Siebenaler, Study Director Ms. Margo Francesco, Administrative Supervisor
PMCD/USACDRA POC:	Ms. Suzanne Fournier, Public Affairs Specialist Ms. Donna Shandle, Director, CSEPP Mr. Tim Thomas, Project Manager, TOCDF

10:00 a.m.–5:00 p.m.	*Tooele County Courthouse*
10:00–11:00 a.m.	**CHEMICAL DEMILITARIZATION CITIZENS ADVISORY COMMISSION**
MG John L. Matthews Chairman, (USA Retired)	Introductions
Dr. Suzanne Winters State Science Advisor	Tooele Advisory Commission, a history
MG John L. Matthews Chairman, (USA Retired)	Program Issues • CSEPP Concerns • Risk Assessment • Alternative Technologies • Allegations by Mr. Steve Jones • M-55 Rocket Stability
All	Discussions with Commission Members

APPENDIX D

11:00–11:45 a.m.	**UTAH DIVISION OF COMPREHENSIVE EMERGENCY MANAGEMENT (CEM)**

Mr. Don Cobb, Chief
CEM Natural and Technological
Hazards Bureau

CEM Introductions and Welcoming Remarks

Utah Chemical Stockpile Emergency
Preparedness Program (CSEPP) Mission

Utah "T.E.E.M. C.S.E.P.P." Concept
- Multi-jurisdictional Approach
- Focus on Teamwork

Utah CSEPP Functional Area Update
- Planning
- Exercise
- Reentry/Restoration
- Training
- Public Affairs
- Health/Medical
- Automation/Communications

Utah CSEPP Readiness: Critical First-Responder Issues
- Personal Protective Equipment (PPE)
- First Responder Operations Focus
- Planning/Training Exercise
- Monitoring

Utah CSEPP Jurisdictional Comments
- Tooele County
- Utah County
- Salt Lake County
- Tooele Army Depot

Questions and Answers

11:45 a.m.–12:30 p.m.   **UTAH DEPARTMENT OF ENVIRONMENTAL QUALITY**

Mr. Dennis R. Downs
Executive Director

1:30–2:00 p.m.   **VISIT TOOELE COUNTY EMERGENCY OPERATIONS CENTER**

2:00–5:00 p.m.   **CITIZENS MEETING**

LETTERS OF INVITATION

Citizens Advisory Commission Invitation Letter

March 7, 1995

MG John L. Matthews, USA Retired
Chairman
Chemical Demilitarization Citizens Advisory Commission
Governor's Office of Planning and Budget, Room 116
State Capitol Building
Salt Lake City, Utah 84114

Dear General Matthews:

For more than seven years, the National Research Council's Committee on Review and Evaluation of the Army Chemical Stockpile Disposal Program (Stockpile Committee) has been providing technical analysis and guidance to the U.S. Army regarding its program of research, development, construction, and operations relating to the task of eliminating the nation's stockpile of lethal unitary chemical agents and munitions. As Chairman of the Stockpile Committee, I am writing to inform you of an information gathering meeting on community concerns regarding the Army's Chemical Stockpile Disposal Program. This meeting will be hosted by members of the Stockpile Committee at Tooele, Utah, on March 29, 1995. The committee's first such community meeting was held on January 4, 1995, at Aberdeen and Kent Counties in Maryland. It was quite useful and informative.

The Committee will be seeking information about various facets of the Army's Chemical Stockpile Disposal Program (CSDP), including such aspects as:

- the concerns of the community as they relate to the implementation of the CSDP;
- the opportunities and mechanisms for community involvement in the CSDP;
- the nature of community involvement in and the status of the Chemical Stockpile Emergency Preparedness Program (CSEPP); and
- other issues of concern to affected parties.

The Utah Citizens Advisory Commission's perspective regarding the CSDP is important to the Stockpile Committee. During its upcoming visit to Utah, the committee would appreciate the commission providing a briefing on this perspective relating to the aspects listed above, and on the commission's responsibilities regarding the CSDP. The committee has set aside time in the meeting agenda from 10:00-11:00 a.m. on March 29 at the Tooele City Hall, 47 South Main Street, for the commission's presentation and any ensuing discussion. Please extend an invitation to all members of the commission to attend. Should you accept this invitation, you may coordinate your presentation with Mr. Donald L. Siebenaler of the National Research Council staff in Washington, D.C., at (202) 334-2577.

Thank you for your interest and efforts on this most important local and national issue.

Sincerely yours,

Richard S. Magee, Chairman
Committee on Review and Evaluation of the
Army Chemical Stockpile Disposal Program

cc: The Honorable Michael O. Leavitt
The Honorable Robert F. Bennett
The Honorable Orrin G. Hatch
The Honorable James V. Hansen
The Honorable William Orton
The Honorable Enid G. Waldholtz
Mr. Dennis R. Downs *(Division of Solid and Hazardous Waste;*
   Intergovernmental Consultation and Coordination Board)
Ms. Lorayne Frank *(Division of Comprehensive Emergency*
   Management; Intergovernmental Consultation & Coordination Board)

APPENDIX D

Public Invitation Letter

March 7, 1995

Mr. John Doe
123 Main Street
Anywhere, USA 00000

Dear Mr. Doe:

For more than seven years, the National Research Council's Committee on Review and Evaluation of the Army Chemical Stockpile Disposal Program (Stockpile Committee) has been providing technical advice and counsel to the U.S. Army regarding its program of research, development, construction, and operations relating to its task to eliminate the nation's stockpile of lethal unitary chemical agents and munitions. As the Chairman of the Stockpile Committee, I am writing to inform you of an information gathering meeting planned by members of the Stockpile Committee at Tooele, Utah, on March 29, 1995. The committee's first such meeting was held on January 4, 1995, at Aberdeen and Kent County communities in Maryland. It was quite useful and informative.

The Committee is seeking information about various facets of the Army's Chemical Stockpile Disposal Program (CSDP), including such aspects as:

- the concerns of the community as they relate to the implementation of the CSDP;
- the opportunities and mechanisms for community involvement in the CSDP;
- the nature of community involvement in and the status of the Chemical Stockpile Emergency Preparedness Program (CSEPP); and
- other issues of concern to affected parties.

Your perspectives and suggestions regarding the CSDP are important to the Stockpile Committee. Specifically, the committee requests written comments on any of the aspects listed above no later than April 24, 1995. Please send them to Mr. Donald L. Siebenaler of the National Research Council staff at the following address:

> Board on Army Science and Technology
> Room HA 258
> National Research Council
> 2101 Constitution Avenue, N.W.
> Washington, D.C. 20418

Additionally, during the committee's March 29 meeting at the Tooele City Hall, 47 South Main Street, several committee members will have limited time to hear summary comments from interested parties on the CSDP between the hours of 2-5 p.m. Should you or your representative wish to address the committee, we have allocated approximately five minutes for each presentation. Enclosed is a response form where you may indicate your choice of time for meeting with the committee members. This form should be returned no later than March 22, 1995. You or your representative will then be contacted and provided an approximate time to address the committee.

For more information, please contact Mr. Siebenaler (202) 334-2577 or Ms. Margo Francesco (202) 334-1902 at the National Research Council. Thank you for your interest and efforts on this most important local and national issue.

Sincerely yours,

Richard S. Magee, Chairman
Committee on Review and Evaluation of
Army Chemical Stockpile Disposal Program

Enclosures: Green Response Form
             Utah Distribution List

DISTRIBUTION LIST

COMMITTEE ON REVIEW AND EVALUATION OF THE ARMY CHEMICAL STOCKPILE DISPOSAL PROGRAM

CITIZENS MEETING

WEDNESDAY, MARCH 29, 1995
Tooele City Hall, 47 South Main Street, Tooele, Utah

Briefings Requested from:

Mr. Dennis Downs
Executive Director
Utah Department of Environmental Quality
Division of Solid and Hazardous Waste
Salt Lake City, Utah

Ms. Lorayne Frank
Public Safety Department
Comprehensive Emergency Management Division
Salt Lake City, Utah

MG John L. Matthews, USA Retired
Chairman
Chemical Demilitarization Citizens Advisory Commission
State Capitol Building
Salt Lake City, Utah

STATE OF UTAH MAILING LIST

Governor Michael O. Leavitt
State Capitol
Salt Lake City, Utah

Honorable Eli H. Anderson
State Representative
Tremonton, Utah

Honorable Robert F. Bennett
U.S. Senate
Washington, DC

Honorable James Gowans
State Representative
Tooele, Utah

Honorable James V. Hansen
U.S. House of Representatives
Washington, DC

Honorable Orrin G. Hatch
U.S. Senate
Washington, DC

Honorable William Orton
U.S. House of Representatives
Washington, DC

Honorable Enid Waldholtz
U.S. House of Representatives
Washington, DC

Mr. Doug Ahlstrom
County Attorney
Tooele, Utah

Ms. Anne Allred
Erda, Utah

Mr. Scott Anderson
Branch Manager
Division of Solid and Hazardous Waste
Salt Lake City, Utah

Ms. Linda Armington
Director
Tooele County Public Health
Tooele, Utah

Dr. William Banner
Division of Pediatric Critical Care
Salt Lake City, Utah

Mr. Malcolm Beck
Provo, Utah

Ms. Relky Bell
Tooele, Utah

Mr. Rex Benmon
Tooele, Utah

SGT David Bennett
CSEPP Coordinator
Utah County Division of Emergency Management
Provo, Utah

Dr. S. John Bennett
Thiokol Corporation
Brigham City, Utah

Mr. Glade Berry
Lehi, Utah

Mr. E. James Bradley
Salt Lake County Commission
Salt Lake City, Utah

Colonel Jesse Brokenburr
Commander
Tooele Army Depot
Tooele, Utah

Honorable Cosetta Castagno
Mayor of the Town of Vernon
Vernon, Utah

Mr. David Clark
Stansbury Park, Utah

Mr. Edward Coale
Systems Manager for Tooele Test Operations
Tooele, Utah

Mr. Don Cobb
Bureau Chief
State Office Building
Salt Lake City, Utah

Ms. Janet Cook
Grantsville, Utah

APPENDIX D

Mr. Leo Coonradt
CSEPP Program Coordinator
State Office Building
Salt Lake City, Utah

Mr. Dave Daniels
Salt Lake City, Utah

Councilwoman Coleen DeLaMare
Tooele City Council
Tooele, Utah

Honorable George Diehl
Mayor
Tooele, Utah

Mr. Ron Elton
Tooele County Attorney
Tooele, Utah

Mr. Steven Erickson
Utah Issues
Salt Lake City, Utah

Councilman David Faddis
Tooele City Council Chairman
Tooele, Utah

Mr. D. Fifield
SARA Title III Program Manager
State Office Building
Salt Lake City, Utah

Mr. Mike Ford
Science and Technology Corporation
Tooele, Utah

Mr. Bipin Gandhi
Ammo Equipment Directorate
Tooele Army Depot
Tooele, Utah

Mr. Gary Griffith
Tooele County Commissioner
Tooele, Utah

Mr. Randy Hall
Tooele, Utah

Ms. Mary Hammong
Grantsville, Utah

Mr. Gary Herbert
Provo, Utah

Commissioner Rancy Horluchi
Salt Lake County Commission
Salt Lake City, Utah

Sidney Hullinger
McFarland Hullinger Co.
Tooele, Utah

Chairman Teryl Hunsaker
Tooele County Commissioner
Tooele, Utah

Mr. Wendell Jensen
Cedar Fort, Utah

Mr. Richard Johnson
Provo, Utah

Mr. Troy Johnson
Grantsville, Utah

Mr. Leo Kelland
T&E Coordinator
State of Utah, Division of CEM
State Office Building
Salt Lake City, Utah

Ms. Corrine Kenney
Utah CEM
State Office Building
Salt Lake City, Utah

Ms. Cindy King
Sierra Club Representative
on the L.E.P.C.
Salt Lake City, Utah

Dr. Richard Koehn
Vice President of Research
University of Utah
Salt Lake City, Utah

Mr. Allen Leung
Salt Lake City, Utah

Dr. Eugene Loh
Department of Physics
University of Utah
Salt Lake City, Utah

Dr. James MacMahon
Dean
College of Science
Utah State University
Logan, Utah

Honorable George Mantes
Tooele, Utah

Mr. Brad Maulding
Division of Solid and Hazardous Waste
Salt Lake City, Utah

Commissioner Lois McArthur
Tooele County Commissioner
Tooele, Utah

Captain Ray McKaye
Utah Highway Patrol
Salt Lake City, Utah

Ms. Norma Miner
Tooele, Utah

Honorable Brenda Morgan
Mayor of the City of Wendover
Wendover, Utah

Honorable Howard L. Murray
Mayor of the City of Grantsville
Grantsville, Utah

Honorable Ray Nelson
Mayor of the City of Stockton
Stockton, Utah

BG David Nydam
Salt Lake City, Utah

Councilwoman Karen Oldroyd
Tooele City Council
Tooele, Utah

Mr. David Ostler
Salt Lake City, Utah

Commissioner Brent Overson
Salt Lake County Commission
Salt Lake City, Utah

Mayor Grant "Bud" Pendleton
Tooele, Utah

Chief Jess Peterson
Tooele Police Department
Tooele, Utah

Councilman Don Peterson
Tooele City Council
Tooele, Utah

Mr. Elwood Powell
Salt Lake City, Utah

Sheriff Donald Proctor
Tooele County Sheriff's Office
Tooele, Utah

Mr. John Ready
Salt Lake City, Utah

Mr. Mark Roberson
Salt Lake City, Utah

Honorable Odell Russell
Mayor of Rush Valley City
Rush Valley, Utah

Ms. Marianne Rutishauser
Tooele County CSEPP Manager
Tooele, Utah

Mr. Doug Sagem
Tooele, Utah

Ms. Kari Sagers
Director
Tooele County Emergency
Management
Tooele, Utah

Mr. Jim Salmon
Division of Solid and Hazardous Waste
Salt Lake City, Utah

Sheriff Frank Scharmann
Tooele, Utah

Ms. Rachel Shilton
Division of Solid and Hazardous Waste
Salt Lake City, Utah

Honorable Walter G. Shubert
Mayor of the City of Ophir
Tooele, Utah

Dr. Paul Skyles
Superintendent of Schools
Tooele School District
Tooele, Utah

Mr. Robert Smith
Resident Engineer
U.S. Army Corps of Engineers
Tooele, Utah

Mr. Ed St. Clair
Tooele County Commissioner
Tooele County Courthouse
47 South Main Street
Tooele, Utah

Mr. Dennis S. Stanley
L.E.P.C. Chairman
Salt Lake County Fire/
Emergency Services
Salt Lake City, Utah

Mr. Gary Swan
Tooele, Utah

Mr. Robert Swensen
Salt Lake County
Emergency Services
Salt Lake City, Utah

Ms. Vicki Varela
Office of the Governor
State Capitol
Salt Lake City, Utah

Mr. Jim Wangsgard
Division of Solid and
Hazardous Waste
Salt Lake City, Utah

Mr. Everett Ward
Tooele County Clean
Air Coalition
Grantsville, Utah

Ms. Beverly White
Tooele, Utah

Councilman Roy Whitehouse
Tooele City Council
Tooele, Utah

Dr. William G. Wilson
Vice President
Hercules, Inc.
Magna, Utah

Dr. Suzanne Winters
State Science Advisor
Office of Planning and Budget
State Capitol
Salt Lake City, Utah

Mr. David Yarborough
Stockton, Utah

Ms. Dorothy D.S. Yu
EG&G Defense Materials, Inc.
Tooele, Utah

Ms. Elizabeth Zimmerman
Utah County Emergency Management
State Office Building
Salt Lake City, Utah

E
Biographical Sketches

Dr. Richard S. Magee, *chair* is a professor in the Department of Mechanical Engineering and the Department of Chemical Engineering, Chemistry, and Environmental Science and is executive director of the Center for Environmental Engineering and Science at New Jersey Institute of Technology (NJIT). He also directs U.S. Environmental Protection Agency's Northeast Hazardous Substance Research Center as well as the Hazardous Substance Management Research Center, which is jointly sponsored by the National Science Foundation and the New Jersey Commission on Science and Technology, both headquartered at NJIT. He is a fellow of the American Society of Mechanical Engineers (ASME) and a diplomate of the American Academy of Environmental Engineers. Dr. Magee's research expertise is in combustion, with major interest in the incineration of municipal and industrial wastes. He has served as vice chairman of the ASME Research Committee on Industrial and Municipal Wastes and as a member of the United Nations Special Commission (under Security Council Resolution 687) Advisory Panel on Destruction of Iraq's Chemical Weapons Capabilities. He presently serves as a member of the North Atlantic Treaty Organization Science Committee's Priority Area Panel on disarmament technologies.

Dr. Elisabeth M. Drake, *vice chair*, a member of the National Academy of Engineering, is the associate director of the Massachusetts Institute of Technology Energy Laboratory. A chemical engineer with interest and experience in technology associated with the transport, processing, storage, and disposal of hazardous materials, as well as with chemical engineering process design and control systems, she has a special interest in the interactions between technology and the environment. Dr. Drake has served extensively as both a consultant to government and industry and as a professor of chemical engineering. She has been very active with the American Institute of Chemical Engineers, in particular with their Center for Chemical Process Safety. She belongs to a number of environmental organizations, including the Audubon Society, the Sierra Club, and Greenpeace.

Dr. Dennis C. Bley is president of Buttonwood Consulting, Inc., and a principal of The WreathWood Group, a joint venture supporting multidisciplinary research in human reliability. He has more than 25 years of experience in nuclear and electrical engineering, reliability and availability analysis, plant and human modeling for risk assessment, diagnostic system development, and technical management. He began his career in 1968 as an officer in the Navy's nuclear reactor engineering program, after graduating from the Massachusetts Institute of Technology. He is a registered professional engineer in the State of California. Dr. Bley has served on a number of technical review panels for Nuclear Regulatory Commission and Department of Energy programs and is a frequent lecturer in short courses for universities, industries, and government agencies. Active in many professional organizations, he holds office in the Institute of Electrical and Electronic Engineers, the Society for Risk Assessment, the Orange County Engineering Council, and the International Association for Probabilistic Safety Assessment and Management. He has published extensively on subjects related to risk assessment. Dr. Bley's current research interests include bringing risk analysis to diverse technological systems, modeling uncertainties in risk analysis and risk management, technical risk communication, and human reliability analysis.

Dr. Colin G. Drury is currently a professor of industrial engineering at the State University of New York at Buffalo and executive director at the Center for Industrial Effectiveness. He has served in a number of professional capacities including committees of the National Institute of Occupational Safety and Health and the National Institutes of Health. His expertise is in human factors and ergonomics, and he has written numerous publications on human factors.

Mr. Gene H. Dyer was graduated with a bachelor of science degree in chemistry, mathematics, and physics from the University of Nebraska. Over a 12-year period he worked for General Electric as a process engineer, the U.S. Navy as a research and development project engineer, and the U.S. Atomic Energy Commission as a project engineer. He then began a more than 20-year career with the Bechtel Corporation in 1963. First a consultant on advanced nuclear power plants and later a program supervisor for nuclear facilities, he then served as manager of the Process and Environmental Department from 1969 to 1983. This department provided engineering services related to research and development projects, including technology probes, environmental assessment, air pollution control, water pollution control, process development, nuclear fuel process development, and regional planning. He culminated his career at Bechtel by serving as a senior staff consultant for several years, with responsibility for identifying and evaluating new technologies and managing their further development and testing for practical applications. He is a member of the American Institute of Chemical Engineers and is a registered professional engineer. He recently served as a member of the National Research Council (NRC) Committee on Alternative Chemical Demilitarization Technologies.

Major General Vincent E. Falter spent more than 34 years in the Army, about half of that time dealing with nuclear weapons. He was Director of Nuclear and Chemical Warfare on the Army Staff and was the single point of contact for all chemical operations for the Department of Defense. He was then responsible for all chemical weapons and their destruction. He initiated funding for the Johnston Atoll Chemical Agent Disposal System and testified on behalf of the system before Congress. He retired from the Army approximately five years ago. Since then, he has been a national security research analyst and consultant for numerous corporations. He has participated in a number of activities, including (a) Joint Strategic Targeting Planning Staff at the Strategic Air Command; (b) Scientific Advisory Committee for Nuclear Weapons Effects; and (c) Department of Defense negotiator for two of the rules for chemical disarmament talks.

Dr. Ann Fisher, senior research associate, Department of Agricultural Economics and Rural Sociology, The Pennsylvania State University, has extensive academic experience. She also spent 10 years at the Environmental Protection Agency, where she analyzed the benefits of reducing environmental risks and then managed the Risk Communication Program. She initiated the Risk Communication Specialty Group within the Society for Risk Analysis. Her research examines how people form perceptions of risk and how those perceptions (and related behavior) change when new information is provided.

Dr. J. Robert Gibson is the assistant director of the Haskell Laboratory, E.I. du Pont de Nemours & Company, and an adjunct associate professor of marine studies at the University of Delaware. After receiving his Ph.D. in physiology from Mississippi State University, Dr. Gibson specialized in toxicology for more than 20 years. Certified by the American Board of Toxicology, he has written numerous publications.

Dr. Charles E. Kolb is president and chief executive officer of Aerodyne Research, Inc. At Aerodyne since 1971, his principal research interests have included atmospheric chemistry, combustion chemistry, chemical lasers, gas/surface methods for advanced materials preparation, and the chemical physics of rocket and aircraft exhaust plumes. He has served on several National Aeronautics and Space Administration panels dealing with ozone in the atmosphere, as well as on two NRC committees dealing with atmospheric chemistry.

Dr. David S. Kosson was graduated with a bachelor of science degree in chemical engineering, a master's degree in chemical and biochemical engineering, and a doctorate in chemical and biochemical engineering from Rutgers–The State University of New Jersey. He joined the faculty at Rutgers in 1986 and was made an associate professor with tenure in 1990. He teaches graduate and undergraduate chemical engineering courses. In addition, he is the projects manager for the Department of Chemical and Biochemical Engineering, where considerable work is under way in developing microbial, chemical, and physical treatment methods for hazardous waste. He is responsible for project planning and coordination, from basic research through full-scale design and implementation. Dr. Kosson is a participant in several Environmental Protection Agency advisory panels involved in waste research and is the director of the Physical Treatment Division of the Hazardous Substances Management Research Center in New Jersey. He is a prolific writer in the fields of chemical engineering and waste management and treatment. He is a member of the American Institute of Chemical Engineers. He recently served as a member of the NRC Committee on Alternative Chemical Demilitarization Technologies.

Dr. Walter G. May was graduated with a bachelor of science degree in chemical engineering and master of science degree in chemistry from the University of Saskatchewan and with a doctor of science degree in chemical engineering from the Massachusetts Institute of Technology. He joined the faculty of the University of Saskatchewan as a professor of chemical engineering in 1943. In 1948, he began a distinguished career with Exxon Research and Engineering Company, where he was a senior science advisor from 1976 to 1983. He was professor of chemical engineering at the University of Illinois from 1983 until his retirement in 1991. There he conducted courses in process design, thermodynamics, chemical reactor design, separation processes, and industrial chemistry and stoichiometry. Dr. May has published extensively, served on the editorial boards of *Chemical Engineering Reviews* and *Chemical Engineering Progress*, and has obtained numerous patents in his field. He is a member of the National Academy of Engineering and a fellow of the American Institute of Chemical Engineers, and he has received special awards from the American Institute of Chemical Engineers and the American Society of Mechanical Engineers. He has a particular interest in separations research work. He is a registered professional engineer in the state of Illinois. He recently served as a member of the NRC Committee on Alternative Chemical Demilitarization Technologies.

Dr. Alvin H. Mushkatel, professor of public affairs, School of Public Affairs, and director, Office of Hazards Studies, Arizona State University, is an expert in emergency response and communications. His research interests include emergency management, natural and technological hazards policy, and environmental policy. He has been a member of the NRC Committee on Earthquake Engineering. His most recent research focuses on the intergovernmental policy conflicts involving high-level nuclear waste disposal and the role of citizens in this policy area.

Mr. Peter J. Niemiec, a partner in the law firm of Greenberg, Glusker, Fields, Claman & Machtinger, in Los Angeles, is an expert in environmental law and regulations. His work in the private sector has focused on the regulation of, and liability arising out of, hazardous materials, including extensive work on Superfund issues. Mr. Niemiec has also represented federal and state environmental agencies, where he was involved in the development of national enforcement policies, and permitting and enforcement issues for major industrial facilities and landfill disposal sites. Mr. Niemiec currently serves as a vice chair of the American Bar Association Special Committee on Toxic and Environmental Torts. He also served as an adjunct professor at the Indiana School of Law (Indianapolis), where he taught environmental law. He has published several articles on the availability of private remedies for environmental cleanup.

Dr. George W. Parshall is a member of the National Academy of Sciences; has been with the Central Research Department of E.I. du Pont de Nemours & Company for nearly 40 years, including 13 years as director–chemical science; and is an expert in conducting and supervising chemical research, particularly in the area of catalysis and inorganic chemistry. He is a past member of the NRC Board on Chemical Science and Technology and has played an active role in National Research Council and National Science Foundation activities.

Dr. James R. Wild was graduated with a bachelor of arts degree from the University of California, Davis, and with a doctorate in cell biology from the University of California, Riverside. Following service as a research microbiologist-biochemist at the U.S. Navy Medical Research Institute, he joined the faculty at Texas A&M University in 1975 as an assistant professor of genetics. He was associate professor of biochemistry and genetics from 1980 to 1984 and was appointed professor of biochemistry and genetics in 1984. In addition to being an extremely active teacher, he has served the university in various administrative positions: currently as chairman of the Faculty of Genetics, professor and head of the Department of Biochemistry and Biophysics from 1986 to 1990, and executive associate dean/associate dean for academic programs of the College of Agriculture and Life Sciences from 1988 to 1992. Dr. Wild has conducted and directed extensive genetic and biochemical research and has published more than 70 scientific articles and participated in countless seminars and invited presentations. He has been a member of the Faculty of Toxicology and has held an NIEHS Graduate Student/Postdoctoral Training Grant in Toxicology since 1992. He recently served as a member of the NRC Committee on Alternative Chemical Demilitarization Technologies.

Dr. Jya-Syin Wu, principal and senior engineer of Advanced System Concepts Associates (ASCA), holds a Ph.D. in nuclear science and engineering from the University of California, Los Angeles. Early in her career she was an associate scientist at the Institute of Nuclear Energy Research in Taiwan, where she held considerable responsibilities in the development of probabilistic risk assessments for nuclear power plants throughout that country. With ASCA since 1991, she has broad experience with probabilistic risk assessments; system reliability analyses; development and application of models for software safety, reliability, and quality assurance; and development and application of expert systems, automated reasoning, and advanced software techniques for automated process management of complex engineering systems.

References

Amos, C.N. 1994. Tooele Chemical Demilitarization Facility Quantitative Risk Assessment Task Overview. Presentation to the Utah Citizens Advisory Commission, July 27, 1994.

Baronian, C. 1994. Correspondence from Charles Baronian, Program Manager for the Chemical Demilitarization, to Dr. Suzanne Winters, Chairman, Utah Citizens Advisory Commission, January 20, 1994.

Benjamin, J.R. 1994. Probabilistic Seismic Hazard Assessment for the TOCDF. Prepared for Science Applications International Corporation by Jack R. Benjamin and Associates and Geomatrix Corporation, December 1, 1994.

Brandyberry, M.D., and C.N. Amos. 1994. Tooele Chemical Demilitarization Facility Quantitative Risk Assessment and Risk Management Plan. Prepared for the U.S. Army Chemical Materiel Destruction Agency by Science Applications International Corporation, April 12, 1994.

Bradbury, J., K. Branch, J. Heerwagen, and E. Liebow. 1994. Community Viewpoints of the Chemical Disposal Program. Washington, D.C.: Battelle Pacific Northwest Laboratories.

Busbee, W.G. 1994. Memorandum from Brig. Gen. Walter G. Busbee to Alma Moore, Principal Deputy Assistant Secretary of the Army, September 30, 1994.

Cobb, D. 1995. Division Perception of the Chemical Stockpile Disposal Program. Briefing by Mr. Don Cobb, Utah Division of Comprehensive Emergency Management, presented to the Committee on Review and Evaluation of the Army Chemical Stockpile Disposal Program at a public meeting, Tooele, Utah, March 29, 1995.

Downs, D.R. 1994. Correspondence from Mr. Dennis R. Downs, director, Utah Division of Solid and Hazardous Waste, to Colonel Jesse Brokenburr, commanding officer, Tooele Army Depot, and Mr. Tim Thomas, project manager, Tooele Chemical Agent Disposal Facility, October 24, 1994.

Downs, D.R. 1995. Division Perception of the Chemical Stockpile Disposal Program. Briefing by Mr. Dennis R. Downs, Utah Division of Solid and Hazardous Waste, presented to the Committee on Review and Evaluation of the Army Chemical Stockpile Disposal Program at a public meeting, Tooele, Utah, March 29, 1995.

EG&G (Edgerton, Germerhausen and Grier Defense Materials, Incorporated). 1994a. Tooele Chemical Agent Disposal Facility, Phase 3 Systemization Demonstration Report, Brine Reduction Area Lines 1 and 2, August 26, 1994. Tooele, Utah: EG&G.

EG&G. 1994b. Dunnage (DUN) Furnace System Thermal Capacity Performance Test Report. Tooele, Utah: EG&G.

EG&G. 1994c. Dunnage (DUN) Furnace System Water Spray Purge Test Report. Tooele, Utah: EG&G.

EG&G. 1995a. Draft Environmental Compliance Plan. Tooele, Utah: EG&G.

EG&G. 1995b. Subject Facility Construction Certification Report Support Documentation for the National Research Council. Tooele, Utah: EG&G.

FEMA (Federal Emergency Management Agency) and Department of the Army. 1994. Planning Guidance for the Chemical Stockpile Emergency Preparedness Program. Washington, D.C., July, 1994.

Fournier, S. 1995. Briefing by Suzanne Fournier to the Committee on Review and Evaluation of the Army Chemical Stockpile Disposal Program, Public Affairs Office, U.S. Army Chemical Demilitarization and Remediation Activity (USACDRA), March 30, 1995.

GPS Technology, Incorporated. 1993. Bulk Drain Station Drain Verification System, GPA-02430, November 10, 1993.

Hance, B.J., Chess, and P.M. Sandman, 1988. Improving Dialogue with Communities: A Risk Communication Manual for Government. Environmental Communication Research Program, New Brunswick, N.J: Rutgers University Press.

Holmes, R.A. 1995. Areas Where the QRA Has Affected Operational Approaches, AMCPM-CDO-O, Memorandum, June 22, 1995.

Insight Research, 1994. Public Opinion Research Data Results. Tooele County, Utah: Tooele County Department of Emergency Management.

Leavitt, M.O. 1993. Letter from Michael O. Leavitt, governor of Utah, to Colonel Nolan Adams, Office of the Assistant Secretary of the Army (Installations, Logistics and Environment).

Lee, M. 1995. Telephone conversation between Myron Lee, Public Information Officer, Tooele County Department of Emergency Management, and a member of the Committee on Review and Evaluation of the Army Chemical Stockpile Disposal Program, August 10, 1995.

McCall, S. 1995. Telephone conversation between Shawn McCall, Tooele County Department of Emergency Management, and a member of the Committee on Review and Evaluation of the Army Chemical Stockpile Disposal Program, April 18, 1995.

MITRE (MITRE Corporation). 1991. Evaluation of the GB Rocket Campaign: Johnston Atoll Chemical Agent Disposal System: Operational Verification Testing. McLean, Virginia: MITRE Corporation.

MITRE. 1992. Evaluation of the VX Rocket Test: Johnston Atoll Chemical Agent Disposal System: Operational Verification Testing. McLean, Virginia: MITRE Corporation.

MITRE. 1993a. Evaluation of the HD Ton Container Test: Johnston Atoll Chemical Agent Disposal System: Operational Verification Testing (April). McLean, Virginia: MITRE Corporation.

MITRE. 1993b. Evaluation of the HD Projectile Test: Johnston Atoll Chemical Agent Disposal System: Operational Verification Testing (May). McLean, Virginia: MITRE Corporation.

MITRE. 1993c. Summary Evaluation of the Johnston Atoll Chemical Agent Disposal System: Operational Verification Testing (May). McLean, Virginia: MITRE Corporation.

MITRE. 1994. Assessment of CSDP Munitions Tracking Capability (Final Draft—MTR 94W0000106–November). McLean, Virginia: MITRE Corporation.

MITRE, 1995. Assessment of CSDP Munitions Tracking Capability: Modified Robotic Munitions Transfer System (May); McLean, Virginia: MITRE Corporation.

Moynihan, P.I., L.E. Compton, J. Houseman, J.J. Kalvinskas, and J.B. Stephens. 1983. Safe Disposal Techniques for DoD Toxic Waste. Volume I, Final Technical Report for the Period 17 June 1982 to 17 August 1983. JPL D-918. Pasadena, California: Jet Propulsion Laboratory.

NAS (National Academy of Sciences). 1969. Report of the Disposal Hazards of Certain Chemical Warfare Agents and Munitions. Ad Hoc Advisory Committee of the National Academy of Sciences. Washington, D.C.: National Academy Press.

National Defense Authorization Act for Fiscal Year 1993. 1992. S. 3114. Congressional Record, 138:(129), September 21.

NRC (National Research Council). 1984. Disposal of Chemical Munitions and Agents, National Research Council. Committee on Demilitarizing Chemical Munitions and Agents. Washington, D.C.: National Academy Press.

NRC. 1993a. Letter report to Assistant Secretary of the Army to recommend specific actions to further enhance the CSDP risk management process. Washington, D.C.: Board on Army Science and Technology, January 8, 1993.

NRC. 1993b. Evaluation of the Johnston Atoll Chemical Agent Disposal System Operational Verification Testing: Part I. Letter report to the Assistant Secretary of the Army. Washington, D.C.: Board on Army Science and Technology.

NRC. 1993c. Alternative Technologies for the Destruction of Chemical Agents and Munitions. Committee on Alternative Chemical Demilitarization Technologies, National Research Council. Washington, D.C.: National Academy Press.

NRC. 1994a. Evaluation of the Johnston Atoll Chemical Agent Disposal System Operational Verification Testing: Part II. Committee on Review and Evaluation of the Army Chemical Stockpile Disposal Program, National Research Council. Washington, D.C.: National Academy Press.

NRC. 1994b. Review of Monitoring Activities Within the Army Chemical Stockpile Disposal Program. Committee on Review and Evaluation of the Army Chemical Stockpile Disposal Program, National Research Council. Washington, D.C.: National Academy Press.

NRC. 1994c. Recommendations for the Disposal of Chemical Agents and Munitions. Committee on Review and Evaluation of the Army Chemical Stockpile Disposal Program, National Research Council. Washington, D.C.: National Academy Press.

OTA (Office of Technology Assessment), U.S. Congress. 1992. Disposal of Chemical Weapons: An

Analysis of Alternatives to Incineration. Washington, D.C.: U.S. Government Printing Office.

Parsons (Ralph M. Parsons Co.). 1993. System Hazard Analysis for the Tooele Chemical Agent Disposal Facility: vols. 1 and 2. Pasadena, California: The Ralph M. Parsons Company.

Parsons, 1994. TOCDF NO_x Data Analysis and Correction of Mass and Energy Balances (December). Pasadena, California: The Ralph M. Parsons Company.

Rasmussen, 1975. Reactor Safety Study. U.S. Nuclear Regulatory Commission, WASH-1400 (NUREG-75/0147).

Rutishauser, M. 1995a. Letter from Marianne Rutishauser, Chemical Stockpile Emergency Preparedness Program (CSEPP) Manager, Tooele County Department of Emergency Management, to Ms. Donna Shandle, Director, Chemical Stockpile Emergency Preparedness Program. Available from Program Manager for Chemical Demilitarization, Aberdeen Proving Ground, Maryland.

Rutishauser, M. 1995b. Telephone conversation with Marianne Rutishauser, CSEP Manager, Tooele County Division of Emergency Management, April 18, 1995.

Sagers, K. 1995a. Statement made by Kari Sagers to the Stockpile Committee at a public meeting, Tooele, Utah, March 29, 1995.

Sagers, K. 1995b. Statement to Procurement Subcommittee of the House Committee on National Security hearings, July 13, 1995.

SAIC (Science Applications International Corporation). 1995a. Tooele Chemical Agent Disposal Facility Quantitative Risk Assessment: Campaigns 1 and 2. Final Draft, June 26, 1995.

SAIC. 1995b. Literature Search and Market Survey of Near Real-time Monitoring Technologies. Report prepared for the U.S. Army Chemical Demilitarization and Remediation Activity, March, 1995.

SAIC. 1995c. Worker Accident Rates at the TOCDF and JACADS. Briefing by the TOCDF safety officer to the Committee on Review and Evaluation of the Army Chemical Stockpile Disposal Program at Tooele, Utah, March 30, 1995.

Shandle, D. 1995. Briefing by Donna Shandle, Director, Chemical Stockpile Emergency Preparedness Program, to the Committee on Review and Evaluation of the Army Chemical Stockpile Disposal Program at Washington, D.C., January 5, 1995.

St. Pierre, G. 1994. Correspondence from Gregory St. Pierre, U.S. Army Chemical Demilitarization and Remediation Activity, to Dr. Suzanne Winters, chairman, Utah Citizens Advisory Commission, August 4, 1994.

St. Pierre, G. 1995a. Information paper regarding Tooele Chemical Demilitarization Facility Risk Management Task Deliverable Schedule. Presented to Committee on Review and Evaluation of the Army Chemical Stockpile Disposal Program during visit to the Tooele Chemical Agent Disposal Facility, June 15, 1995.

St. Pierre, G. 1995b. Information paper regarding the Stockpile Contingency Action Plan. Presented to Committee on Review and Evaluation of the Army Chemical Stockpile Disposal Program during visit to the Tooele Chemical Agent Disposal Facility, June 15, 1995.

St. Pierre, G. 1995c. Information paper regarding an Agricultural Risk Assessment. Presented to Committee on Review and Evaluation of the Army Chemical Stockpile Disposal Program during visit to the Tooele Chemical Agent Disposal Facility, June 15, 1995.

Taylor, J.M. 1994. Proposed Policy Statement on the Use of Probabilistic Risk Assessment Methods in Nuclear Regulatory Activities, SECY-94-218, August 18, 1994.

Tooele County, 1994a. Tooele County Emergency Operations Plans. Tooele, Utah: Tooele County Department of Emergency Management.

Tooele County, 1994b. Communications Plan (draft). Tooele, Utah. Tooele County Department of Emergency Management, October 17, 1994.

U.S. Army. 1974. Chemical Agent Data Sheets, Vol. 1. Technical Report, EO-SR-74001 (Edgewood Arsenal Special Report). Alexandria, Virginia: Defense Technical Information Center.

U.S. Army. 1987. Chemical Stockpile Disposal Program, Risk Analysis of the Disposal of Chemical Munitions at Regional or National Sites. SAPEO-CDE-IS-87008. Aberdeen Proving Ground, Maryland: U.S. Army.

U.S. Army. 1988. Chemical Stockpile Disposal Program Final Programmatic Environmental Impact Statement (PEIS). Available from Program Manager for Chemical Demilitarization, Aberdeen Proving Ground, Maryland.

U.S. Army. 1993. Required Report for the Operational Verification Tests (part of TOCDF RCRA Permit—October 1993). Available from Program Manager for

Chemical Demilitarization, Aberdeen Proving Ground, Maryland..

U.S. Army. 1993–1995. TOCDF Functional Analysis Workbook. Available from Program Manager for Chemical Demilitarization, Aberdeen Proving Ground, Maryland.

U.S. Army. 1994a. Courtesy Chemical Surety Inspection—Tooele Chemical Agent Disposal Facility (6 September 1994); Washington, D.C.: Office of the Inspector General.

U.S. Army. 1994b. Safety Evaluation Report (22 November 1994). Washington, D.C.: Army Safety Office.

U.S. Army. 1994c. Tooele Chemical Agent Disposal Facility—Report on Design-Related Safety Issues and Evaluation of Construction Conformance with Design (9 December 1994). Huntsville, Alabama: Engineer Division.

U.S. Army, 1994d. Tooele Chemical Agent Disposal Facility Campaigns 1 and 2 Quantitative Risk Assessment (draft), May 9, 1994. Available from Program Manager for Chemical Demilitarization, Aberdeen Proving Ground, Maryland.

U.S. Army. 1994e. Tooele Chemical Agent Disposal Facility Quantitative Risk Assessment Methodology Manual, December 22, 1994. Available from Program Manager for Chemical Demilitarization, Aberdeen Proving Ground, Maryland.

U.S. Army. 1995a. JACADS Brine Reduction Area Mass and Energy Balance, EPA Demonstration Test Report, vol. 1. Available from Program Manager for Chemical Demilitarization, Aberdeen Proving Ground, Maryland.

U.S. Army. 1995b. Trial Burn Report for Agent GB in the Dunnage Incinerator at JACADS, EPA Demonstration Test Report, vol. 1. Available from Program Manager for Chemical Demilitarization, Aberdeen Proving Ground, Maryland.

U.S. Army. 1995c. TOCDF Risk Management Plan (draft, April 1995). Available from Program Manager for Chemical Demilitarization, Aberdeen Proving Ground, Maryland.

U.S. Army. 1995d. Final Protocol, RCRA (Resource Conservation and Recovery Act), Part B. Risk Assessment No. 73710536-39-1399-95, Anniston Chemical Demilitarization Facility, U.S. Army Center for Health Promotion and Preventive Medicine, February 6, 1995.

U.S. Army and Air Force, 1975. Military Chemistry and Chemical Compounds. Field Manual No. 3-9, Air Force Regulation No. 355-7. (October 30). Washington, D.C.: U.S. Department of the Army Headquarters.

USATHAMA (U.S. Army Toxic and Hazardous Materials Agency). 1982. Long Range Chemical Demilitarization Concept Study—Revised. Available from Program Manager for Chemical Demilitarization, Aberdeen Proving Ground, Maryland.

U.S. Code of Federal Regulations. 1976. Resource Conservation and Recovery Act; 40 CFR Part 264; Washington, D.C.: Environmental Protection Agency.

U.S. General Accounting Office. 1993a. Chemical Weapons Destruction: Issues Affecting Program Cost, Schedule, and Performance. GAO/NSIAD-93-50. Washington, D.C.: U.S. Government Printing Office.

U.S. General Accounting Office. 1993b. Chemical Weapons Storage: Communities Are Not Prepared to Respond to Emergencies. GAO/NSIAD-93-91. Washington, D.C.: U.S. Government Printing Office.

U.S. General Accounting Office. 1994. Chemical Weapons Stockpile: Army's Emergency Preparedness Program Has Been Slow to Achieve Results. GAO/NSIAD-94-91. Washington, D.C.: U.S. Government Printing Office.

U.S. General Accounting Office. 1995. Chemical Weapons: Army's Emergency Preparedness Program Has Financial Management Weakness. GAO/NSIAD-95-94. Washington, D.C.: U.S. Government Printing Office.

U.S. Nuclear Regulatory Commission. 1994. Probabilistic Risk Assessment Implementation Plan Public Workshop, Rockville, Maryland, December 2, 1994.